A PLUME BOOK

POINCARÉ'S PRIZE

GEORGE G. SZPIRO (GeorgeSzpiro.com) was born in 1950 in Vienna, Austria, and later moved to Zurich, Switzerland. He studied mathematics at the Eidgenössische Technische Hochschule (ETH) in Zurich, obtained an M.B.A. at Stanford University, and did his Ph.D. in mathematical economics at the Hebrew University in Jerusalem. Szpiro has been assistant professor of finance and decision sciences at the Wharton School of the University of Pennsylvania, and then at the Hebrew University. He has written thirty scientific papers in mathematics, physics, economics, and finance.

He became an Israel correspondent and science writer for the Swiss daily *Neue Zürcher Zeitung* and the weekly *NZZ am Sonntag*. He has been honored by the Swiss Academy of Sciences with the Prix Média, was a finalist for the Descartes Prize in Science Communication (awarded by the European Commission), and was a recipient of the Media Prize from the German Association of Mathematicians (DMV).

His previous books are *Kepler's Conjecture* and *The Secret Life of Numbers*.

George lives with his family in Jerusalem.

POINCARÉ'S

Prize

The
Hundred-Year
Quest to Solve
One of
Math's Greatest
Puzzles

George G. Szpiro

A PLUME BOOK

PLUME
Published by the Penguin Group
Penguin Group (USA) Inc., 375 Hudson Street, New York, New York 10014, U.S.A. • Penguin Group
(Canada), 90 Eglinton Avenue East, Suite 700, Toronto, Ontario, Canada M4P 2Y3 (a division of
Pearson Penguin Canada Inc.) • Penguin Books Ltd., 80 Strand, London WC2R 0RL, England • Penguin
Ireland, 25 St. Stephen's Green, Dublin 2, Ireland (a division of Penguin Books Ltd.) • Penguin Group
(Australia), 250 Camberwell Road, Camberwell, Victoria 3124, Australia (a division of Pearson Australia
Group Pty. Ltd.) • Penguin Books India Pvt. Ltd., 11 Community Centre, Panchsheel Park, New Delhi – 110
017, India • Penguin Group (NZ), 67 Apollo Drive, Rosedale, North Shore 0632, New Zealand (a division
of Pearson New Zealand Ltd.) • Penguin Books (South Africa) (Pty.) Ltd., 24 Sturdee Avenue, Rosebank,
Johannesburg 2196, South Africa

Penguin Books Ltd., Registered Offices: 80 Strand, London WC2R 0RL, England

Published by Plume, a member of Penguin Group (USA) Inc. Previously published in a Dutton edition.

First Plume Printing, August 2008
3 5 7 9 10 8 6 4 2

Mathematical editing: Christina Sormani, Lehman College and CUNY Graduate Center

 REGISTERED TRADEMARK—MARCA REGISTRADA

The Library of Congress has catalogued the Dutton edition as follows:
Szpiro, George, 1950–
Poincaré's prize : the hundred-year quest to solve one of math's greatest puzzles / George Szpiro.
p. cm.
Includes bibliographical references and index.
ISBN 978-0-525-95024-0 (hc.)
ISBN 978-0-452-28964-2 (pbk.)
1. Mathematics—Problems, exercises, etc. 2. Mathematics—Popular works. I. Title.
QA43.S985 2007
510.76—dc22   2007012792

Printed in the United States of America
Original hardcover design by Leonard Telesca

*Dedicated to Grisha Perelman, a man of rare genius and modesty.
His devotion to mathematics and his refusal to let glory get
the better of him are admirable.*

Contents

Chapter 1

Fit for a King

A ugust is not the best time of year to visit Spain. Soaring temperatures during the late summer make even short sightseeing trips and brief walks through medieval cities a plight. But tourism was not the reason that several thousand men and women from all over the world flocked to Madrid on August 22, 2006. Their motivation for traveling to Spain from 143 countries and all continents except Antarctica was the 25th International Congress of Mathematicians. Commonly known as the ICM, this gathering is a grand affair that takes place only once every four years. The previous ICM had been held in Beijing, the one before in Berlin. Four thousand mathematicians from around the world, among them the most distinguished professors from the most prestigious universities and research institutes, had gathered in Spain this time to attend twenty plenary sessions and more than a thousand invited lectures, short talks, and poster sessions. Over ten days, scientific meetings alternated with concerts, exhibitions, and cultural activities. The ICM's main event was the opening ceremony, at which Juan Carlos I, the king of Spain, awarded the prestigious Fields Medals to four mathematicians.

Already two hours before the official opening at ten A.M., a long queue of people wound around the modern Palacio Municipal de Congresos in the northeastern part of the Spanish capital. The expected presence of the king necessitated strict security measures, and each entrant was searched before being given access to one of the conference halls. While the thousands of participants whiled away the time in a long human snake outside the conference center, the air was abuzz with speculations about who would be honored with a Fields Medal. Considered the equivalent of the Nobel Prize, these medals are the highest honor the mathematics profession has to offer. Even though the monetary reward is only a tiny fraction of the Nobel Prize's, the Fields Medals are arguably even more select than the Nobel Prizes, with at most four winners chosen every four years. The identities of the medalists are a closely kept secret up until the very last moment. To encourage talent, the rules specify that recipients must not be older than forty years at the start of the year in which the ICM takes place.

Over many months preceding the ICM, a secret committee consisting of nine of the world's top mathematicians had screened the work of dozens of brilliant candidates. Finally, at the end of May, they had chosen the four best from among them. Those who knew the results were under strict orders not to divulge them. The awardees themselves had been informed during the summer, but they were not to tell anybody of their impending honor, except possibly their spouses. Not even they knew the identities of the other medalists, however.

To the mathematical world, a mysterious Russian from St. Petersburg was the most deserving of the impending honor. Grigori Perelman, the hands-down favorite in the queue outside the conference center, had all the right qualifications: He had proven his worth with sensational papers in the preceding years, some of them unpublished. He was just two months into his fortieth year, his birthday having fallen on the previous June 13. But most important, he had solved one of the oldest unsolved problems of mathematics: the 102-year-old Poincaré's Conjecture. This problem was considered of such fundamental importance by the community of mathematicians that when the former Fields Medalist Steve Smale was asked in 1998 to name the most important mathematical problems for the coming century, he put Poincaré's Conjecture near the top of his list. And the Clay Mathematics Institute in Boston listed it as one of the seven most mind-boggling unsolved math problems anyone on earth could imagine.

But on this day in August, Perelman was nowhere to be found. In fact, he spent the festive day hidden away in the modest apartment that he shared with his mother in a drab neighborhood of St. Petersburg. Perelman had no interest whatsoever in the Fields Medal. If the king of Spain had hoped to meet the Russian mathematician, he had come in vain.

This book will tell the story of the Poincaré Conjecture and of the attempts to solve it, from the times before it was even formulated until the moment that "Grisha" Perelman made it into a theorem.

The ICM is held under the auspices of the International Mathematical Union. The first such conference took place in Zurich, Switzerland, in 1897, the second one in Paris in 1900. Since then the event has been held every four years, except for gaps during and after the two World Wars. In contrast to the hundreds of specialized mathematics conferences that take place every year, the all-encompassing ICMs give all the members of the profession an opportunity to mingle. Experts from different fields who would probably never meet one another or learn about one another's work can interact. The meetings grew from a modest start— 204 men and 4 women attended the Zurich meet over a century ago—to a gigantic happening a century later.

The Fields Medals are a more recent invention. The brainchild of the otherwise undistinguished Canadian mathematician John Charles Fields, who died in 1932, they were first bestowed on two mathematicians, Lars Ahlfors from Finland and Jesse Douglas from the United States, in 1936. Due to World War II, the awards were discontinued for the next fourteen years but resumed in 1950. Since then, between two and four Fields Medals have been awarded every four years. Altogether, until Madrid, forty-four men—no women as yet—have been awarded this highest honor of the mathematicians' guild. Never had anybody turned it down.

In 2006 the charge of organizing the ICM fell upon the president of the International Mathematical Union, the Englishman Sir John Ball from Oxford University. Ball was also chairman of the Fields Medal Committee, and as such he knew the identities of the prizewinners: Terence Tao from UCLA, a thirty-one-year-old former Australian child prodigy of Chinese descent; Andrei Okounkov, a Russian who taught at Princeton University; the German-born Frenchman Wendelin Werner from the Université Paris-Sud; and yes, indeed, Grigori Perelman from...well, that wasn't quite clear. Having resigned from the Steklov Institute of Mathematics of the Russian Academy of Sciences in St. Petersburg a few months ago, Perelman was unaffiliated with any institution.

Three years previously, he had briefly visited the United States to present his proof at MIT, Princeton, SUNY, NYU, and Columbia. He had been unkempt, humble, and brilliant. Everybody who attended his lectures was deeply impressed, and lucrative job offers followed. But Perelman was not interested. Immediately following his last lecture, he returned to St. Petersburg. For a while afterward he still answered e-mail queries by colleagues about his proof, but when he felt that his methods had been understood, he stopped corresponding.

Well aware of Perelman's idiosyncratic personality and fearing that the awards ceremony in Madrid would not go as planned, Sir John traveled to St. Petersburg in mid-June in order to meet this person with the mysterious reputation. Sir John had set himself the task of informing Perelman of the impending honor and confirming that he would attend the awards ceremony in Madrid. A public meeting between Perelman and the president of the International Mathematical Union would have given the secret of the Fields Medal away, so the two first met in a secluded location to talk. Ball soon realized that his fears had been justified. During the next two days, he tried to convince Perelman to follow the august union's customs and play by the rules. Perelman was pleasant enough and gave him a long walking tour of his native city. But on the main subject he remained adamant. He did not want the Fields Medal.

Making no headway and believing that it was the publicity that Perelman shunned, Ball offered him a different option. Accept the honor from afar and the medal would be sent at a later date. But Perelman would have none of that either. He had no interest in honors, nor did he want any public recognition. Proving the Poincaré Conjecture was award in itself. Frustrated and disappointed, Ball left Russia. But he accepted Perelman's right to decline the award. "He has a different psychological makeup, which makes him see life differently," Ball would later tell the journalists assembled in Madrid.

But there was more to Perelman's inexplicable behavior than that. Apparently he was trying to make a point. Manuel de León, the Spanish professor who was serving as chairman of the ICM, had talked to Perelman himself. "The reason he gave me," he said, "is that he feels isolated from the mathematical community and therefore has no wish to appear as one of its leaders." And Ball hinted ominously that "certain personal experiences with the mathematical community during his career had caused him to remain at a distance." Unfortunately, Ball was not at liberty to disclose the nature of these experiences, he explained. To this day, nobody apart from John Ball and possibly some of Perelman's close confidants knows what these experiences were. Whatever they may be, it is clear that Perelman wishes to remain above the inevitable compromises of political and social life inside and outside academia.

So if he does not care about honor and recognition, maybe money could entice him? The incentive was there. The Clay Mathematics Institute in Boston, created in 1998 by the businessman Landon T. Clay and his wife, Lavinia D. Clay, to "increase and disseminate mathematical knowledge," offers a million-dollar prize for the solution of any of their seven Millennium Problems. As noted above, Poincaré's Conjecture is one of them. For Perelman the prize was, and still is, up for grabs. All he needs to do is to stretch out and reach for it. Not surprisingly, the

Russian mathematician made no attempts to fulfill the requirement for the prize, which was to publish his proof in a reputable mathematical journal. He made do with the posting of three spectacular papers on the Internet and left it at that, million dollars or no million dollars.

One person who wanted to know more about Perelman's intentions was the author Sylvia Nasar. Having previously written a best seller about the mathematician John Nash—who had suffered from paranoid schizophrenia, been in and out of mental institutions, and then went on to win the Nobel Prize in economics—she was curious about this strange mathematician. Together with the science journalist David Gruber, she found Perelman to be cordial and frank. They also found a man who was so idealistic as to seem out of touch with the real world.

According to a former colleague from the Steklov Institute who prefers to remain anonymous, Perelman is so deeply disappointed by the perceived decline of ethics in the mathematics community that he no longer considers himself a professional mathematician. Some regret about this did show through in his conversation with Nasar and Gruber. "It is not people who break ethical standards who are regarded as aliens," he told them, "it is people like me who are isolated."

Why Perelman resigned from the Steklov Institute in December 2005 is shrouded in mystery, but he had not detached himself completely from the mathematics community even after his resignation. The Fields Medal forced him to make a clean break. "As long as I was not conspicuous, I had a choice," he told Nasar, "either to make some ugly thing"— presumably to speak out about the perceived lax standards in the mathematics community—"or, if I didn't do this kind of thing, to be treated as a pet. Now, when I became a very conspicuous person, I cannot stay a pet and say nothing. This is why I had to quit." His criticism did not spare anyone. "Even those who are more or less honest tolerate those who are not."

There may be another reason for Perelman's severance of relations with the mathematics community. The history of science is rife with priority disputes, and mathematics is no exception. Perelman's reluctance to publish his proof made it easy for others to make a grab for the glory by capitalizing on his success. Perelman professes not to bear a grudge. "I can't say I'm outraged. Other people do worse," he commented in reference to such an attempt. But true to his character he abhors controversies, and rather than get involved in disputes that seemed inevitable, he apparently preferred to cut all ties to his former colleagues.

Perelman's absence left a big void in Madrid, but the ICM went ahead as planned. At a scientific meeting whose organizing committee is known to be extremely selective about which subjects are to be presented to the plenum, no less than three lectures dealt with Poincaré's Conjecture and Perelman's proof. The verdict was unanimous. The teams that had taken it upon themselves to study Perelman's Internet postings agreed that the papers contained no error and constituted a complete proof of the century-old conjecture. In fact, it turned out Perelman's proof went well beyond Poincaré's Conjecture.

In mathematics there is no such thing as a final stamp of approval. Unlike the Académie Française, which can sanction a new word in the French language, or the International Astronomical Union, which can decide whether Pluto is, or is not, a planet, no central authority exists that can vet a theorem. Acceptance of a mathematical proof is a gradual process governed by customs and traditions. But the pronouncements in Madrid came as close to an official endorsement as was possible. ICM 2006 will be remembered as the event at which Poincaré's Conjecture finally became a theorem. One of the seven hardest mathematical problems of our age had at last been solved.

Chapter 2

What Flies Know and Ants Don't

Standing at the command bridge of his converted cargo ship *Santa Maria*, Christopher Columbus faced a problem. His men were close to a mutiny. It was October 9, 1492, and the little fleet consisting of the caravels *Niña* and *Pinta* in addition to the *Santa Maria* had been sailing on the Atlantic Ocean for over two months. The crew—just under a hundred officers, sailors, deckhands, stewards, carpenters—had became restive. They were scared. If the voyage continued much longer, the ships would get too close to the edge of the world and eventually fall over the brink. The earth was flat, after all, and nobody knew what abyss lay beyond its borders.

Most of Columbus's more educated contemporaries back home actually knew that the earth was round. Otherwise they would surely not have been prepared to fund Columbus's expedition. A voyage to India going westward instead of toward the east makes sense only if the earth is a sphere. (Actually, a cylinder would also do.) In fact, the ancient Greeks had already known that the earth was a sphere. Watching ships arrive in the harbor, they noticed that, at a distance of about fifteen kilometers, a boat with a ten-meter-high mast is completely hidden below the horizon. As it draws nearer, the top of the mast comes into view, and little by little the lower parts of the boat become visible. To the astute Greek thinkers this was an indication that the earth was a sphere.

The philosopher Eratosthenes (276–194 BC), most famous for his prime-number sieve, had even calculated the earth's circumference. By planting vertical poles into the ground at the towns of Aswan and Alexandria and observing their shadows, he inferred that the circumference of the earth must have the length of 250,000 stadia. The only remaining question: How long is a stadium? Scholars now estimate its length at between 157 and 166 meters, which puts the circumference of the earth, as measured by Eratosthenes, at between 39,250 and 41,500 kilometers. With the true circumference being 40,080 kilometers, as determined by satellites today, Eratosthenes' estimates carried an astonishingly small error of no more than 2 to 4 percent.

Compare this with Columbus's error seventeen centuries later. He had underestimated the earth's circumference by a full third. Actually, we may be giving Eratosthenes too much credit; his measurements were not that accurate. The distance between Alexandria and Aswan was not known exactly, the latter city was not exactly south of the former, the length of the stadium is today unknown, the times of measurement could not be synchronized exactly. However, most of the errors seem to have somehow canceled one another out.

Steeped in medieval beliefs and superstitions, the men on the three Spanish boats were not educated. All their experiences and voyages told them that the earth was flat. To conceive of anything else would have stretched credulity. Fortunately, just three days later, on October 12, 1492, land was sighted. The mutiny was averted and Columbus had discovered America—albeit without knowing it. He thought that he had circled half the globe and arrived in India.

But let us not be too patronizing toward the poor sailors on the *Santa Maria*, the *Niña*, and the *Pinta*. After all, in the Middle Ages and the Renaissance it did take quite a leap of faith to acknowledge that the earth is a sphere. Forget the Renaissance, even today there are stick-in-the-mud diehards who believe that the earth is a flat disk. Members of the Flat Earth Society believe that the north pole is in the center of a disk and the edges consist of insurmountable, fifty-meter-high ice mountains.

They are convinced that the stories of a spherical earth are part of a worldwide conspiracy, moon landings were hoaxes staged by Hollywood, and the space shuttle is an elaborate charade, meant to prop up the dying myth that the earth is a globe. After all, look left, look right, everything is flat. And if antipodes did exist on the other side of a globe, obviously people would have to fall off the earth, since gravity—yes, they are prepared to accept gravity—always pulls downward. Try to beat that argument.

Obviously, a basketball is a sphere. But to a fly or an ant crawling on the ball's surface, the basketball seems completely flat. Actually an ancient Greek, sophisticated in mathematical matters such as, say, Aristotle (384–322 BC), could have ascertained whether the shape on which he walked was flat or curved. He would have attached a piece of rope to a pole, paced out a circle, then measured its circumference. If the circumference was less than 2π times the length of the rope, he would correctly have surmised that the surface he was standing on was curved. But crawling insects, without access to sophisticated measurement methods à la Aristotle, and without the possibility of wandering around the whole object à la Christopher Columbus, would stay ignorant about the true shape of their crawling space.

Actually, many, many shapes look flat in an observer's neighborhood. For example, even Aristotle or Eratosthenes would not have noticed if the earth had a hole through the middle. Without taking careful measurements and traversing the entire surface, insects would notice no difference when crawling on a bagel, on a basketball, or on a flat disk. Even a pretzel would look the same. A fly does have an advantage over the ant, however. It can lift off—into the third dimension—and observe the surface from afar. Then it can realize if the object on whose surface it has been crawling is a basketball or the inner tube of an automobile tire.

Hence, to ascertain whether its crawling space lies on a plane, a sphere, a bagel, or a pretzel, insects have to take a step back—to lift off— and look at the object at arm's length. But only flies can do that. This may seem clear enough, but trust mathematicians to challenge even the obvious. What is the deeper reason for the flies' broader view? The answer is that by flying away from the object, airborne animals are able to move in three dimensions. Ground-based animals, on the other hand, are constrained to crawling around in two dimensions, and that makes all the difference in the world.

Wait a minute, is a basketball not a three-dimensional object? So why are ants constrained to two dimensions? Well, the basketball is a three-dimensional object, but the ant does not move around all of it. It can only wander about the surface of the ball. And the *surface* of a three-dimensional object is two-dimensional. The skin of a ball and the crust of a bagel are two-dimensional objects. The whole of the objects—outer surfaces and all of the inside—is three-dimensional.

That the surface of a sphere like the earth is two-dimensional is also borne out by the fact that Christopher Columbus could determine the position of his ships in the Atlantic Ocean by stating just two coordinates, latitude and longitude. Any place on earth can be determined by two coordinates. The address 7 West Forty-fifth Street pinpoints an exact location on the two-dimensional map of New York City. But the Big Apple is not flat, and a street address may not suffice. Your business partner may be waiting up in her office. In this case three coordinates are needed: 7 West Forty-fifth Street, seventeenth floor. Now the venue of the business meeting is well defined in three-dimensional space. Hence, if a position on or above the face of the earth is to be specified, the value of a third coordinate, say the height above sea level, must be added.

Now that we know that surfaces of three-dimensional objects are two-dimensional, one would like to know what it is, mathematically, that distinguishes ball, bagel, and pretzel from one another. Questions of this sort belong to the discipline of topology. For once mathematicians do not complicate things too much, but only restate the obvious. Ball, bagel, and pretzel differ from one another by the number of holes they have. So what else is new? Maybe this: In topology, all surfaces with the same number of holes are considered identical.

Mathematicians specializing in topology study which bodies can be transformed into one another by pulling, squeezing, twisting, and turning...but not by tearing or gluing. Topologically speaking, a sphere, an egg, and a box are considered the same object because they are holeless. If these objects are made out of clay, each of them can be turned into the other just by deformations. A bagel, a tire, and a coffee cup are one-holed objects and are therefore also considered identical bodies. On the other hand, since a bun cannot be turned into a bagel without punching a hole into the dough, these two pastries are considered different bodies. And a bagel is different from a pretzel, not because one is cinnamon and the other is salty, but because the latter has three holes. Finally, a pretzel is the same object, topologically speaking, as a three-handled teacup: They both have three holes.

Of course, topology would be a bit banal if it limited the objects of its research to just one-, two-, or three-dimensional bodies. Interesting things may happen in higher dimensions, and mathematicians would like to prove general statements that hold true in spaces of any dimension. Even though few people can visualize more than three dimensions, moving from three-dimensional to four-dimensional space often requires no more mental effort than adding one to three. The fourth dimension could be time, for example. An invitation to a meeting at 7 West Forty-fifth Street, seventeenth floor, at ten A.M. would unambiguously fix the venue of the business encounter in the four-dimensional space-time continuum, as would a date at 403 West Fourteenth Street, lower lounge, at ten P.M.

In mathematics, dimensions are a general notion and it is of no importance whether they are identified with addresses and times. What needs to be remembered, though, is that the dimension of an object's surface is one less than the dimension of the object itself. We already pointed out that the surface of the three-dimensional earth is two-dimensional.

In the same manner we may say that the surface of a two-dimensional disk is the one-dimensional circle that surrounds it; the surface of the one-dimensional line is the zero-dimensional endpoint. And—to challenge your imagination—let me point out that the surface of four-dimensional balls are three-dimensional spheres.

Usually a mathematical theorem is first proved for low-dimensional spaces. Once the ice has been broken, one can often build upon the proof to extend the theorem to higher dimensions. It is like adding another floor to a one-story house. If the foundations are sound, this may be feasible. But it is not always an easy process. Take Kepler's Conjecture, which is about how balls can be stacked in the densest manner. In one dimension, balls refer simply to pieces of lines—you could visualize them as matches—and the answer is trivial: Just place them end to end. The two-dimensional case refers to circular shapes such as coins. One easily realizes by trial and error that the way to place coins most densely on a tabletop is to arrange them hexagonally. But the mathematical community had to wait until 1940 for a rigorous mathematical proof that this is, in fact, the densest arrangement of two-dimensional balls. The three-dimensional case refers to balls as we know them. It had always been suspected that the densest arrangement of balls is the one that greengrocers use to stack apples and oranges: in a pyramid. In fact the German astronomer Johannes Kepler already suspected as much in 1611, hence the name of the conjecture. But a proof that this is true was given only in 1998. And the answer for four- and higher-dimensional balls is not even known, let alone proved.

Sometimes it is the other way around. It may be relatively simple to prove a statement in high-dimensional space, while in low dimensions the question remains unanswered. This could happen because the higher-dimensional space provides more elbow room. It is, after all, much easier to store one's belongings on two floors than it is to do so in a single-story house. As we shall see in this book, this is exactly what happened with the Poincaré Conjecture.

But let us not rush things. For the time being, keep in mind that three-dimensional objects are surfaces of four-dimensional bodies. Can we always tell a ball from a bagel by inspecting its surface? In three dimensions it's easy—at least for us and for flies, if not for ants. But would we be able to tell a four-dimensional ball if we tripped over one?

This is where Monsieur Henri Poincaré comes in.

Chapter 3

The Forensic Engineer

Monsieur Henri Poincaré, Professor of Mathematical Astronomy in the University of Paris, Member of the French Academy and of the *Académie des Sciences*, died suddenly this morning of an embolism of the heart. The death of Henri Poincaré at the comparatively early age of 58 deprives the world of one of its most eminent mathematicians and thinkers." Thus reported the correspondent of the *London Times* to its readers on Wednesday, July 17, 1912, from Paris.

And two days later, describing the magnificent funeral, the paper wrote, "The religious ceremony was at the Church of *Saint Jacques du Haut Pas*, whither the coffin had been removed from Monsieur Poincaré's house in the *Rue Claude Bernard*, followed by an illustrious funeral procession of French public men and men of science." The interment took place in the family vault in the cemetery of Montparnasse. Attendees at the burial represented the period's who's who of politics and science. Of course, the prime minister of France was present; he was, after all, the cousin of the deceased. But there were many more. The president of the Senate, ministers, the prince of Monaco, and the bey of Tunis attended. The *président de la république* had sent a personal representative, and among other dignitaries there were delegations from the Académie des Sciences and from the Sorbonne, as well as the president of the Paris Geographical Society, the secretary of the Royal Society of England, and the astronomer royal.

The minister of education, in the funeral oration, called Poincaré "a kind of poet of the infinite, a kind of bard of science," and a member of the Académie Française recalled the glory that the mathematician had brought the fatherland. The correspondent of the *Times* reminded his readers of the great reception that had been accorded Poincaré only a few weeks before his untimely death when he had delivered a series of lectures on higher mathematics at London University.

Who was this exceptional man who was so profusely eulogized? Jules-Henri Poincaré was born in Nancy, a town in the northeast of France, on April 29, 1854. His father, Léon, was a doctor and professor of medicine at the University of Nancy; his uncle, Antoni, inspector of public works and an acknowledged authority on meteorology. Both father and uncle were true intellectuals who not only fulfilled their professional duties with great enthusiasm but also wrote scholarly treatises published by the French Academy of Sciences.

The family can trace its lineage on the father's side back to a certain Jean-Joseph Poincaré, a justice official in the town of Neufchâteau, who died in 1750. There are records of two family members: Aimé-François Poincaré, a soldier who rendered great services to the army during the wars of the Revolution, and a great-uncle, Nicolas-Sigisbert Poincaré, a military commander who fought in the Spanish war under Napoléon and disappeared when the French army retreated in Russia. The latter had adopted the name Pontcarré, and Henri would have preferred this as a surname.

It was for good reason that he never liked his family name. The name phonetically sounds like the French for "square point." Since the times of the ancient Greeks, it was known that a point in its true guise is infinitely small and certainly not square. The misnomer that was his name highly displeased the future mathematician. Pontcarré, on the other hand, means "square bridge," which made a lot more sense. But the secret of the family name goes back further. In the early fifteenth century, a certain Petrus Pugniquadrati was known to have been a student in Paris. Pugniquadrati would translate as "square fight," which would not have been an attractive surname to the rather gentle boy either. But it would at least not have contained any mathematical impropriety. Even earlier, in the late fourteenth century, a certain Jehan Poingquarré had been secretary to Queen Isabella of Bavaria. This name translates as "square fist," which indicated a strong man, and whence Henri's family name probably derives.

Henri's mother was described as good, active, and intelligent, which are just about the highest compliments that could be bestowed on women of the nineteenth century. She devoted all her energies to the education of Henri and of his younger sister, Aline Catherine Eugénie. The precocious boy started to speak at an early age—badly at first, since his mind worked faster than he was able to talk. At age five, he fell seriously ill with diphtheria and needed to recuperate for many months. Afterward he stayed weak and shy, fearing the company of boys of his age. He was even afraid to walk down the stairs on his own. Instead of running around and playing with friends, he took refuge in the world of books, becoming a voracious reader. His photographic memory was legendary, and he was able to cite the exact page and line where he had read a certain passage.

To make up for missed schooling during his illness, his family arranged for a private tutor. The instructor, an exceptional elementary-school teacher and family friend, assigned little homework but encouraged the boy to ask as many questions as he liked. Henri availed himself copiously of this possibility and seemed to be doing all right in his studies until he was ready, at age eight, to enter the regular school system. But since no tabs were kept on his achievements, nobody had any idea what and how much he actually knew. Would he be able to follow the curriculum? The family need not have worried. His very first essay was described as a small masterpiece by his teacher, who ranked the boy first in his class. Henri would keep that rank in all subjects throughout his school years.

Learning came so easily to him that he never seemed to be doing any work. A childhood friend described how Henri would do his school assignments. He would be in his mother's room, walking around, doing all kinds of things, and keeping up a conversation. At various times he would walk by the table and, without even taking a seat, write a few lines into his copybook. Being ambidextrous, he did so with either the right or the left hand. Then he would resume his former activities. After a while, the homework was done, seemingly without any effort at all on the boy's part.

Young Henri had an exceptionally happy childhood. He loved his sister dearly and liked to laugh and play with her and her girlfriends. He was gentle and kind; his intellectual superiority did not make him arrogant. The highlights of the years were the vacations he spent at his grandparents' estate in the countryside. A large garden served as the site for walks and games and provided the grounds for many discoveries. When he was eleven, he was able to explain to his friends the workings of the echo in the mountains, the phenomenon of electricity, and the functioning of the telegraph. At thirteen, he and his friends acted in dramas and comedies that often he had written himself. He liked to dance and was always ready to go out for walks and to amuse himself.

Nothing yet indicated that Poincaré would one day become a great mathematician. His favorite subjects were history and geography, and his aptitude for literary and philosophical essays aroused the interest of his educators. Scribbled on scraps of paper of different formats, his compositions were sometimes considered by his teacher as too daring and original to be formally presented in class. Henri's interest in mathematics only became apparent when he was fourteen years old. While he still successfully pursued his classical studies, his preoccupation with mathematics started to grow. Soon his mathematical abilities became well known among the teaching staff.

There was a dark period in Poincaré's adolescence, however. The Franco-Prussian War broke out when he was sixteen. Old enough to accompany his father on his daily rounds in ambulances and sufficiently mature to help with medical chores and duties, he witnessed some of the horrors of war. Traveling with his mother to her parents' home in the countryside, they came through burned-out and looted villages, empty of all inhabitants. When they finally arrived at the estate where Henri had so loved to play during his childhood summers, they realized with horror that the home had been ransacked. Everything of any value whatsoever had been taken by the Prussians. Even food was scarce. The cellar, the pantry, and the chicken run had been emptied of all contents. The day the enemy left, Henri's grandparents would have gone to bed hungry had it not been for a kind neighbor who brought over a bowl of soup. (This episode actually puts the family's suffering somewhat into perspective. Going to bed hungry was hardly the most gruesome incident the war produced.) The sight of his grandparents' destroyed house and the adolescent boy's experiences during the occupation of his hometown of Nancy kindled in Henri an ardent patriotism that would remain with him for the rest of his life. There was one benefit to the occupation for Poincaré: He learned German in order to read the only newspapers that were available.

In August 1871, Poincaré sat for the literary *baccalauréat*, the French high school exams. He did extremely well in the Latin and French essays and excelled in all other subjects. Three months later, it was time for the mathematical *baccalauréat*. Amazingly, Henri nearly failed the written test. The exam question asked for a proof of a formula to compute the sum of the terms of a geometric series. Henri had arrived late for the examination. Out of breath and slightly nervous, the aspiring student misunderstood the question and botched it. Fortunately, the professors were aware of the examinee's already well-established reputation and agreed to slightly bend the rules. When announcing the names of those who had passed the written exam, the head of the jury proclaimed— when he arrived at Poincaré's name—that "any other student would have been rejected for this mathematical essay." Of course, Poincaré quickly redeemed himself by doing brilliantly in the oral part of the exam.

Following high school, young Frenchmen with exceptional aptitude for mathematics underwent, and still undergo today, a two-year grueling course in mathematics in preparation for entry to the *grandes écoles*, the prestigious engineering schools. (Nowadays women may also compete for admission to the *grandes écoles*.) During this intensive course, Poincaré distinguished himself again. At the end of his first year, he entered a competition in which the best students from all over France participated. He won the first prize.

In the second preparatory year, outstanding students usually rise above the merely very good ones. *Math Spé*, specialized mathematics, deals with subjects that would today be taught in advanced courses at university. Poincaré was unfazed. He hardly took any notes, and when he did once jot something down, it was on a piece of paper that was clearly recognizable by its black border as someone's death notice. At first everybody thought that he had simply forgotten his notebook. But when similar scenes repeated themselves day after day, his colleagues— all of them excellent students—became curious. On the one hand, the newcomer did not seem serious. On the other hand, he had won first prize in the prestigious competition. So what was one to make of him? They decided to subject him to a test. After an exceptionally difficult lecture, during which Poincaré had once again taken hardly any notes, a senior student went up to him and asked him to explain a particularly obscure point. Without hesitation, Poincaré gave a minilecture on the subject. The doubters were left standing, mouths agape. Poincaré's seriousness was never put to question again.

At the end of the second preparatory year, Poincaré again won first prize in a nationwide competition. Following the exam, he had been afraid that he wouldn't even get an honorable mention, since he had written no more than a single page in answer to the prize question. But his contribution was so outstanding that one of the members of the jury later said that he would have placed Poincaré first even if he had made computational errors or left part of his work unfinished, just because of the way he presented his answer. "This student will go far," the professor predicted.

Henri's next challenge was to enter the competition to gain admission to one of France's top engineering schools. Of all the *grandes écoles* the most prestigious was, and is, the École Polytechnique in Paris. The X, as it is called for short—the symbol deriving from the school's military insignia: two crossed cannons—opened its doors in 1794, just after the French Revolution. It had been founded under the name École Centrale des Travaux Publics (Central School of Public Works) by the mathematicians Lazare Carnot and Gaspard Monge. A year later it changed its name to Polytechnique and quickly gained a reputation as a first-class institute of higher learning. In 1798 Napoléon invited forty-two students and professors from the École Polytechnique to accompany him on an expedition to Egypt. Eleven years after its inception, the emperor turned the school into a military academy, whence the crossed cannons, and created for it a motto that managed to combine science with martial-sounding patriotism: *Pour la patrie, les sciences et la gloire* (For fatherland, science, and glory). Two centuries later, in 1970, the Polytechnique was changed to a civilian institute. But it stayed under the tutelage of the Ministry of Defense and kept its pithy motto. Two years later, the first female student was admitted. To the consternation of many a male-chauvinist Frenchman, the young lady was not only allowed to enter the hallowed halls, but did so as the top-ranked incoming candidate.

Nowadays, five hundred students from France and abroad are accepted by the Polytechnique every year following an extremely strict selection process that lasts for several weeks. The rewards are, however, commensurate. Graduates are assured a place in France's scientific, industrial, political, or economic elite. Outstanding scientists, many of whose names are now attached to basic laws in mathematics and fundamental concepts in physics, have walked its corridors as students or professors during the last two centuries. One of the most outstanding was Henri Poincaré.

When Poincaré presented himself for the oral entrance exam, his reputation was already so great that the hall, normally empty during the boring interrogations of students, was standing room only. After receiving his question, Poincaré, deep in thought, talked slowly, sometimes closing his eyes. He seemed to invent mathematical proofs while he was going along. The examiner was nothing less than amazed. The second examiner, who was to test Poincaré's knowledge of geometry, suspended the exam for forty-five minutes before questioning Poincaré so that he could prepare a particularly ingenious question. When the question was posed, the examinee started to pace up and down in front of the blackboard before he suddenly announced, "It all comes down to proving that AB equals CD." "Very good, monsieur," the professor responded, "but I desire a more elementary answer." Poincaré stopped his walking before the professor's table and—out of the blue—orally developed a trigonometric solution to the question. "All right," the examiner challenged, "but I would like you not to stray outside of simple geometry, monsieur." No sooner had he said that than Poincaré gave another proof. This time it was beyond reproach; the professor congratulated him warmly and announced that he had received the highest grade.

Obviously Poincaré was a superb candidate, but there was a snag. His ineptitude at physical exercise—sports is even today an entry requirement for the Polytechnique—was one hindrance. But an even greater impediment was his inability to draw. His grade was zero, and such a mark in any subject made a candidate ineligible for entry to the *grande école*. One of the examiners commented to a colleague in Paris, "In Nancy there is a quite outstanding candidate. But we are embarrassed. If he gets accepted, it will be as the top-ranked candidate. But will he be accepted at all?" To its credit, the school found a way to overcome Poincaré's ambidextrous scribbling, and he was accepted as the top-ranked freshman. In the fall of 1873, Poincaré moved to Paris to attend the École Polytechnique.

As was to be expected, the student did brilliantly in nearly all subjects. He continued his habit of not taking any notes during lectures, and of doing his thinking while walking up and down the school's corridors. But his lack of draftsmanship came to haunt him at graduation, two years later. The French love to place just about everyone and everything in order of merit, and after their ranking of the *grandes écoles* and the incoming students, the young men must, of course, be ranked again at the end of their studies. Poincaré towered above his fellow students throughout his studies in all subjects except those that required freehand drawing. His inability to draw straight lines and maybe his tendency to leave out explanations in his proofs because everything seemed so obvious to him angered an examiner and earned him a mediocre grade in geometry. As a result, Poincaré finished as only the second-ranked graduate. (The year's *majeur*—first-ranked graduate—was a certain Marcel Bonnefoy, who died tragically at age twenty-seven. As an engineer, he was charged with the investigation of an accident in the coal mine of Champagnac in which a worker had died. Shortly after he entered the mine, a gas explosion occurred, immediately killing Bonnefoy and two miners. Two more engineers died of their wounds a few days later.)

Poincaré continued his training with three years of graduate studies at the École des Mines. Founded in 1783, this *grande école* is even older than the Polytechnique. In the eighteenth and nineteenth centuries the industrial revolution was in full swing, and mining for raw materials was considered quite high-tech. The school's charter was to train young men as senior engineers in the service of the state. The curriculum comprised subjects that were of importance to mining operations, like mineralogy, geology, railway construction, mining legislation, agriculture. Some more general courses also had to be taken, English or German, physical fitness, and…drawing. Admission to the school was and is based on, you guessed it, a competition. Depending on their rank at graduation, *polytechniciens* who have already acquired the scientific foundations are admitted straight to the advanced program of the Corps des Mines. This program trains future managers of public and private entities. They receive the technical, economic, and social tools to be active at the interface between state and enterprise for the good of the nation. Nowadays, about two-thirds of the fifteen or so yearly entrants to the Corps des Mines are *polytechniciens*.

Poincaré entered the École des Mines together with Bonnefoy and another *polytechnicien* by the name of Petitdidier. While Poincaré pursued his training as a future engineer and manager, his interests started to turn more and more toward science. But he had little time to pursue his interests. In vain, a former teacher from the Polytechnique tried to get Poincaré exempted from some of the courses so that he would have more time to further his mathematical knowledge. Poincaré had to follow the full curriculum. After all, the École des Mines considered itself a *grande école* as well and would not take any flak from the competition.

The French school and university system evaluates the accomplishment of students with grades between zero and twenty. Twelve is the passing threshold. To interpret the French grading system one must know, however, that grades 18 to 20 are unavailable for mere mortals, 16 to 18 are reserved for the professor himself, and anything above 13 is considered outstanding. While high grades were practically nonexistent, on the low end, grades were calculated to the second place after the decimal as befits an engineering school. In addition, an intricate system of balancing the relative importance of the subjects and a weighing of each year's marks gives the final grade.

With this in mind, let us look at Poincaré's achievements at École des Mines. In exploitation (of mines, not of workers) and machinery, the subject given the most weight at École des Mines, Poincaré obtained a grade of 17.17; in paleontology he got 17.32 and in mineralogy 17.40. The latter grade is especially noteworthy, since the professor, François-Ernest Mallard, routinely failed even very good first-year students. (Poincaré's two colleagues from the Polytechnique didn't make it. They had to retake the exam a year later.) All his other grades were good to very good, except for English and drawing, which he failed. Second year, same story, except that this time he passed English with a very respectable 15.41. His drawing abilities had also improved somewhat: With a grade of 11.99, he failed by just one-hundredth of a point. No rounding at the École des Mines; this is, after all, a school with a reputation for extreme precision. Given that his grade point average was 15.10, or "well beyond 12," as his report card said, Poincaré was allowed to enter the final year.

Luckily, freehand drawing was no longer a required subject and the student could breathe a sigh of relief. Instead there were field trips. First there was a geological excursion to Normandy, about which Poincaré sent enthusiastic letters to his family back home. But the École des Mines did not limit its extramural activities to France. At a time when voyages abroad were arduous and rare, the school already wanted its students to gain international experience. Poincaré traveled twice beyond the French borders at the school's behest: once to Hungary and a year later to Scandinavia. After his first trip he wrote reports about coal mining on the estates of the Austrian *Staatsbahn* (state railway lines), and about the metallurgy of tin in the Hungarian region of Banat. His second trip resulted in two reports about the exploitation of mines in Scandinavia.

Unfortunately, the four reports have not been found in the archives of the École des Mines, but apparently his professors were happy with them. At the school board's meeting of June 13, 1878, Poincaré was declared a graduate of the École des Mines. The twenty-four-year old engineer and freshly minted member of the Corps des Mines was ready to assume responsibilities.

Graduation from the École des Mines is only the beginning of a career as a mining engineer. By a decree of December 24, 1851, the French Republic had established a hierarchy of advancement at the *service des mines*. The lowest rung, just a notch above the students, was the *ingénieur ordinaire*. Next in line was the *ingénieur en chef*, and finally, at the top of the pecking order, was the *inspecteur général*. Each rank was further subdivided into two or three classes. Advancement from one class to the next, and from one rank to the next, was possible after a minimum of two years on the job at the previous level. Thus the top spot, *inspecteur général première classe*, could be attained only after about fifteen to twenty years of service, provided there was still an opening, because their number was limited to at most three.

Salaries varied accordingly. While an *ingénieur ordinaire* third class received 1,800 francs yearly, the *inspecteur général* first class got 12,000 francs in addition to office and traveling expenses. In comparison, university professors made between 1,500 and 2,000 francs—just about as much as an ordinary engineer third class—a captain at sea took home 2,700 francs, and a professor at a military academy made a cool 3,000 francs. As behooves any self-respecting civil service, job security went above all else. The president of the republic himself needed to pronounce the firing of an engineer. On the other hand, an engineer could not resign without the president's consent.

On March 28, 1879, Poincaré was named *ingénieur ordinaire* and was set for a distinguished career in the *service des mines*. A week later, he received his first posting, as supervising engineer of the coal mines in the region of Vesoul in eastern France near the German and Swiss borders. He was delighted with this posting because Vesoul was close to his family home.

During the next eight months the novice engineer visited the mines in the region under his supervision no less than five times. In June he traveled to Saint-Charles, whose mines were already nearly empty of coal deposits. Poincaré noted with precision their irregular and poor performance. In September 1879, he went to the mine at Magny to investigate a tragic incident. Only a few weeks later he visited the mine at Saint-Pauline, where his interest centered on air ventilation, exhaust of gases, and water seepage. In October he inspected the cast-iron lining around the shafts of the mine of Saint-Joseph. His last visit, again to the mine at Magny, took place during November. The visits to Magny were especially noteworthy, even thrilling, both for the dangers involved and for the nature of Poincaré's work and report.

On September 1, news reached the office of the region's supervising engineer that a massive explosion had occurred in the early-morning hours in the coal mine of Magny. Sixteen people had been killed. As soon as Poincaré was informed of the tragic event, he set out for the site of the accident. He wanted to take a firsthand look immediately. At great danger to his life—the cause of the explosion was still unknown and another one could occur at any moment—he descended into the coal pit to investigate the possible reasons for the accident.

The detailed report that Poincaré wrote during the following three weeks is a model of engineering and forensic research. It would have done the legendary Inspector Maigret, hero of Georges Simenon's best-selling detective novels, credit and gives a taste of how Poincaré worked, connecting leads, considering evidence, rejecting hypotheses, proving his final conclusions. It also revealed something already of the superbly analytical mind that would, twenty-five years later, come up with an apparently simple problem that would remain unsolved for a hundred years.

The mine had only recently started its operations. Coal was being produced from 650 meters below the surface. On August 31, at six o'clock in the evening, twenty-two men entered the pit. It was a Sunday and no work had been carried out during the day. Eleven miners, seven laborers, one mason, and one roadman made up the team. They were led by two foremen instead of the usual one, since one of them was due to leave Magny the next day. He had come one last time to introduce his successor to his new workplace.

At four o'clock in the morning, a strong explosion occurred. All the lights were blown out. Two brothers, Félicien and Amable Miellin, who were at that moment ascending in the cage that carried men to the surface, were violently shaken. They immediately redescended with spare lamps and were able to help four survivors, two of them only slightly wounded, ascend to daylight.

An off-duty master miner by the name of Juif lived close to the mine's exit. He was woken by the explosion, came running, and immediately descended into the mine. Together with the Miellin brothers, he made his way toward the site of the accident. On their way through the narrow galleries, they came across a pile of clothing that had been set alight by the blast. The fire was immediately put out by Juif; otherwise it could have provoked another explosion. At that moment, the men heard the cries of a severely wounded sixteen-year-old laborer, Eugène Jeanroy. He was transported to daylight and straight to a hospital, where he succumbed to his wounds.

Within half an hour of the explosion, two engineers arrived on the scene, accompanied by a master miner and a few workers. Their first task was to repair the ventilation system so that the rescuers could advance farther into the mine. The state of the first bodies that were recovered left little doubt as to the destiny of the other men. There was no hope of finding any survivors. One after another, the bodies were brought to daylight. The last of them could be retrieved only two days after the accident. The master miner Juif was everywhere. His zeal in the rescue efforts was such that he had to be held back by the engineers. The sixty-year-old man stayed in the pit for eighteen straight hours.

Poincaré arrived in Magny in the early afternoon of the same day and immediately descended into the mine to investigate. While exhausted rescuers were rushing about and the danger of a secondary explosion still lurked, he examined the site of the explosion firsthand. Later he sat down to write his report.

Poincaré analyzed all aspects of the mine and its operation. One danger of mining operations is insufficient ventilation. Another is the workers' lamps, which can ignite combustible gases that are often present in a mine. Since the latter posed such a great danger, special care was taken by the mine's management to ensure safety. Miners who did not conform to the strict regulations on the maintenance of their lamps were punished. Poincaré noted that fines of three to five francs had been imposed on thirteen miners in the past for dropping their lamps, piercing them, or twisting their latches. One miner was even dismissed for unscrewing his lamp while at work, and a purveyor was fined ten francs for delivering an unclosed lamp.

A doctor's autopsy reports noted that all sixteen men had burned to death. Since a person located between the entry point of air to a mine and the actual site of an explosion burns to death, while men trapped behind the explosion usually die of asphyxiation, this indicated the location of the explosion. Furthermore, the doctor's reports noted that the soles of the men's feet were undamaged. From this Poincaré concluded that the victims were caught by the accident while standing and had died instantly. At least they were spared the agonies of a slow death, he observed.

This evidence pointed to two possible locations for the explosion. To find out which was the one, Poincaré focused on the lamps that were found inside the mine. They all carried numbers, and the names of the miners to whom they were handed out were always meticulously registered. Lamp number 476 stimulated Poincaré's particular attention. It was found hanging about fifteen centimeters from the ground, in the vicinity of the fire-scarred body of seventeen-year-old laborer Émile Perroz. Its glass was gone and the metal showed two tears. One tear was big and wide and seemed due to pressure from inside the lamp. The other tear aroused Poincaré's suspicion: The metal was bent toward the inside in a little square. This indicated that the shock that had caused the damage must have come from outside the lamp. The evidence indicated a blow as if the lamp had been hit by a pickax.

The damaged lamp might account for what had ignited the gas, but where did the explosive gas, carbonated hydrogen, come from? After inspecting the ventilation system and computing the amount of fresh air that was forced into the mine per second, Poincaré concluded that the mine's aeration was beyond reproach. Any gas gradually seeping into the mine would have been expelled by the ventilation. In addition, just a few hours before the accident, one of the surviving workers had been at the location where Poincaré suspected the explosion had taken place but had not noted any sign of gas. Poincaré posed and dismissed various hypotheses in his report. A gradual seepage of gas into the mine would have had no unfortunate consequences. Had an accumulation of gases taken place in spite of the aeration, the gas would have concentrated first near the ceiling of the corridor. Long before the gas would have reached the lamp—hanging just fifteen centimeters from the ground—and caused an explosion, Perroz would have suffocated from lack of oxygen. The doctor's report showed, however, that the laborer had burned to death. Hence, the reason for the explosion must have been the sudden and rapid escape of a small amount of gas through the lower part of the mine's wall. This also explained the relatively minor physical damage to the mine itself.

One problem remained. Lamp number 476 had been given out to the miner Auguste Pautot, not to Perroz. But its remains were found next to the corpse of Perroz. Furthermore, it was this laborer's job to load the chariots with coal. Hence, he had no pickax. Lastly, Perroz's lamp, number 16, was found intact elsewhere inside the mine.

By now, Poincaré had enough leads to reconstruct, step by step, the events that led to the accident. Apparently, the thirty-three-year-old miner Auguste Pautot had hung his lamp on a piece of wood at his location inside the mine. While working, he must have pierced the lamp with a stroke of his pickax. He did not notice the damage because the fissure, toward the bottom of the lamp, was nearly invisible. The incident was of no consequence at first, because no gas was present where Pautot worked. But for some reason, Pautot then went to where Perroz was working. He hung his lamp next to his colleague's, and when he descended again a while later, took Perroz's lamp by mistake. Pautot's damaged lamp was left hanging, just a step from a potential gas leak. At any other time, the danger might have passed unnoticed, but unfortunately, a sudden escape of gas occurred through the lower part of the mine's wall just when and where the damaged lamp was hanging. The lamp's flame ignited the seeping gas and caused the explosion.

To summarize, Poincaré wrote in his report, the tragic accident was due to the clumsiness of Pautot, who paid for a moment's inattention with his life. The man had not been a bad worker and no complaints had ever been lodged against him. But unfortunately, an imprudence such as this could be committed even by the best miner. The accident killed sixteen men, among them the two foremen.

In spite of his detached tone as a supervising engineer, Poincaré's compassion in the well-researched and detailed report was obvious. He did not put undue blame on the poor Auguste Pautot, whose inattentiveness had caused the accident. Neither did he neglect to mention the personal tragedy of the nine widows, thirty-five children, and two elderly parents who were left without breadwinners. He emphasized the bravery of master miner Juif, who had led the rescue efforts at extreme peril to his own life. The emergency aid of forty francs per family and monthly pensions of twenty-five francs for each widow and eight francs for each orphaned child would not compensate for the human tragedy, Poincaré noted in closing. The only item he omitted in the report was his total disregard for his own personal safety. When he'd first entered the mine, barely twelve hours after the accident, another explosion could have happened at any moment. Just such a secondary explosion would cost Poincaré's colleague Bonnefoy his life two years later. But clarification of the causes of the accident required an immediate investigation, and it was Poincaré's job to undertake it. He followed his duty without hesitation.

Two months later, on November 29, Poincaré traveled to Magny once more. He wanted to see for himself the repair work that had been undertaken since the accident. In particular, the ventilation system, though not faulty, had been upgraded. Poincaré noted with satisfaction that no significant quantities of carbonated hydrogen had been detected since the accident. And whenever small whiffs were noticeable, the gas was expelled through the improved ventilation system within a day or so.

On December 1, 1879, barely eight months after having been named *ingénieur ordinaire* third class, Poincaré wrote his last report. His reputation in mathematics was already great—as will be recounted in the next chapter—and he had just been named professor at the University of Caen. Thus his career as mining engineer ended before it had really started. Poincaré was not overly happy with this turn of events. He would have liked to continue in his profession simultaneously while teaching. But this was not feasible, and he was granted an unlimited leave of absence by the minister of public works. Nevertheless, Poincaré, who was described by those who knew him during that time as superbly competent but modest to the point of timidity, continued to be a member of the Corps des Mines. Advancing through the ranks, he was named *ingénieur en chef* in 1893—albeit without salary or benefits—and *inspecteur général* in 1910.

Many years after leaving active service as a mining engineer, he still occupied himself with mining and the safety of the operations. Mindful of the accident he had investigated more than thirty years earlier, he returned to the art of mining and the associated dangers in one of his last publications. "A spark suffices to ignite the explosive mixture of air and pit gas. I cannot describe the horrors that follow." Far away already from the world of mining and steeped in academic matters, Poincaré still showed how deep his humanity was, how he cared for his fellow men and women.

Poincaré was a formidable, even heroic, engineer. His real calling, however, lay elsewhere.

Chapter 4

An Oscar for the Best Script

Even though Poincaré was an outstanding engineer, his first love remained mathematics. The curriculum at the École Polytechnique, with its classes in mathematics and physics, catered to his tastes. At École des Mines the courses were geared toward technical subjects, and the students were prepared for future management duties. But Poincaré's interest in "the queen of the sciences" never flagged. After all, who could forget the interest that had lead to his winning national prizes as a student?

While studying mineralogy, mining, and other useful subjects at the École des Mines, Poincaré also worked, by himself, on problems of advanced mathematics. At the end of his first year as an engineering student, he had completed a paper on partial differential equations that would later be published by the *Journal de l'École Polytechnique*. One of the readers, the world-famous mathematician Charles Hermite, at the Sorbonne, was impressed but ambivalent. On the one hand, the depth of Poincaré's work was astonishing. On the other hand, the author seemed unable to express his ideas clearly; his style was a little confused and details were omitted. As had happened in his childhood, when little Henri's mind worked faster than he could talk, now too his ideas came to him so rapidly that he did not always have time to work out all the details. When he presented another paper on differential equations to

Jean Darboux, who later became secretary of the Académie des Sciences and would pronounce the eulogy at Poincaré's funeral, the Sorbonne professor extolled the essay's virtues but at the same time strongly criticized its lack of rigor. Poincaré duly ironed out the wrinkles and the paper was presented as his doctoral thesis at the University of Paris. Upon graduation from the École des Mines, it seems Poincaré was already destined for an academic career.

In December 1879—just a few weeks after finishing his investigation of the mining disaster at Magny—the Ministry of Public Works released him from his engineering duties and placed him at the disposal of higher education. By now well-known in mathematical circles, Poincaré was commissioned to teach mathematics at the University of Caen in northwestern France. His somewhat disorganized teaching style was not universally appreciated by the students, but his colleagues, especially his former teacher Hermite, were full of praise. Two years later, Poincaré was appointed to the chair of mathematical physics and probability at the Sorbonne. Simultaneously, he obtained a teaching position at his alma mater, the École Polytechnique. In the same year he married Louise Poulain d'Andecy. The couple had three daughters and a son. Then a challenge presented itself that would make him a household name throughout Europe. The sequence of events did not proceed without some serious snags, however.

Since Johannes Kepler's computations of the orbits of celestial objects and Isaac Newton's discovery of gravitation, nobody had worried whether Earth, Mars, or Venus could one day deviate from their orbits around the Sun and suddenly take a different course. Even today, we consider it obvious that our planetary system remains stable. But maybe we should worry? Could a passing comet cause a commotion someday and throw the planetary system out of kilter?

When Kepler computed the elliptical orbit of Mars, he neglected to notice that on its path around the Sun the celestial body did not exactly describe an ellipse. The observations that he had received, or rather stolen, from his predecessor in Prague, the Imperial Mathematician Tycho

Brahe, were the most exact data available at the time. But even so, they contained minute errors. The orbit is only nearly periodic, or as mathematicians like to say, quasi-periodic. It deviates from a perfect ellipse because a planet's orbit is influenced not solely by the sun's gravitation but also by the gravitation of all the other bodies present in the system. So intrepid scientists—mathematicians, astronomers, physicists—set out to compute the orbits of more than two bodies whose gravitations influence one another simultaneously. Since the two-body problem had been solved by Newton, scientists started with three bodies, thinking that this would not require too much effort. All that had to be done, they thought, was to add a few more equations to the system and solve them. But soon they realized that things were not quite that simple: The so-called three-body problem could not be solved exactly.

When he was fifty-five years old, King Oscar II of Sweden and Norway, a great supporter of science, was wondering how to celebrate his sixtieth birthday, which would fall on January 21, 1889. Gösta Mittag-Leffler, a Swedish mathematician who had rejected an offer of an extraordinary professorship in Berlin to stay in his fatherland and who was now an up-and-coming member of the Swedish academic establishment, had an idea. In his youth, the future king had taken courses at the university and done quite well in mathematics. Later he founded the mathematical journal *Acta Mathematica*, of which Mittag-Leffler became editor in chief. So Mittag-Leffler proposed the establishment of a prize for an essay about the n-body (here n means any number, including 3) problem as an appropriate way to celebrate the king's birthday. With his proposal, he expected to kill three flies with one stroke: The prize would manifest the king's appreciation of science, it would put Scandinavia on the map, and the brouhaha around it would increase Mittag-Leffler's own stature as a mover and shaker in the world of mathematics.

King Oscar liked the idea. If no great theory of mathematics or fundamental law in physics would be attached to his name, at least he could award an Oscar to someone else's work. The monarch charged

Mittag-Leffler with the minutiae of organizing the prize. He was to work out the rules, choose the jury, announce the prize, and finally select the winner.

Mittag-Leffler, who did not yet anticipate how much heartache the project would bring, embarked on it with verve. At first, everything seemed to be going just fine. He was able to persuade two of the most important European mathematicians to serve with him on the jury, his former teachers Charles Hermite in Paris and Karl Weierstrass in Berlin. But problems arose immediately. The choice of Weierstrass deeply offended the irascible Leopold Kronecker, Weierstrass's eternal rival at the University of Berlin. Mittag-Leffler pointed out to him that Weierstrass had only been chosen because he was the older man. This calmed Kronecker down temporarily, but he would continue to pester Mittag-Leffler throughout the competition and during its aftermath, until he died in 1891. Another problem was that Hermite and Weierstrass, even though they were fluent in both German and French, insisted on corresponding in their respective mother tongues. Mittag-Leffler was called upon to serve as an intermediary, translating and forwarding correspondence to and fro. But these problems would be the least of his worries.

To broaden the competition, the members of the jury included three problems. The most interesting was, of course, the n-body problem. The question posed by Weierstrass, slightly reformulated, was "Given a system of n bodies that attract each other according to the law of gravitation, and assuming that no two bodies ever collide, give the coordinates of the individual bodies for any time in the future or the past, as the sum of a uniformly convergent series whose terms are made up of known functions."

The questions and the rules of the competition were duly published in 1885 in *Acta Mathematica*, in *Nature*, and in other publications. The entries had to arrive, anonymously, at the editorial office of *Acta Mathematica* by June 1, 1888. Kronecker immediately criticized first one, then another of the questions. But his criticisms were recognized as the misgivings of a grouch and not taken seriously. All over Europe, mathematicians started working.

By the time the deadline arrived three years later, twelve entries had been received. As the rules stipulated, they were submitted anonymously, identified only by an epigraph on the outside of the envelope. Another envelope, carrying the same epigraph, contained the name and address of the author and would remain sealed until the prize ceremony. To this day, the identities of only four entrants are known. Five essays dealt with the three-body problem. None of them contained a solution to the three-body problem. But the 150-page manuscript that carried the epigraph *nunquam praescriptos transibunt sidera fines* (nothing exceeds the limits of the stars) had achieved enormous advances in the study of the dynamics of moving bodies. So the members of the jury decided that the author of that paper would be the winner of the competition.

On the eve of the king's birthday, the winner was announced in the castle of Stockholm. With great trepidation, the envelope that contained the name was opened in front of a hushed crowd of invited guests: "And the winner of the Oscar for the best script is…Henri Poincaré, for 'The problem of the three bodies.'" In contrast to today's Oscar ceremony, there was no frenetic yelping and hooraying, with no tears shed and no choked-up exclamations of "Thank you, Mom"—just some polite applause while the king signed the protocol.

Poincaré had dealt with the problem by, first of all, restricting it: He assumed a system composed of a large body (like the Sun), a smaller body (like the Earth) and a very small body (represented in our solar system by the Moon). But even the "restricted problem," as it was henceforth called, turned out to be devilishly difficult. Poincaré managed to describe the orbits of the three bodies only approximately as sums of series of numbers. This sufficed as an answer to the prize question, as it had been posed by Weierstrass. But Poincaré did more than that. He proved rigorously that no analytical solutions (i.e., no elegant formulas) exist that would describe the position of the bodies at all times. The surprising implication of this result is that the positions of the planets in our solar system cannot be predicted with total precision.

Despite this bad news, on March 23, 1889, Count de Loewenhaupt, the Swedish ambassador in Paris, awarded Poincaré a gold medal with the king's likeness and the prize money of twenty-five hundred crowns. (To put the prize money in the correct perspective, Mittag-Leffler's annual salary as a professor in Stockholm was seven thousand crowns.) Hermite, who received a copy of the medal in his capacity as judge, albeit in silver, recounted later that his medal had gone from hand to hand during a meeting of the Academy of Sciences and that everybody had been struck by its beauty.

Poincaré inquired about the appropriate manner in which to thank King Oscar. Seeing an occasion to further his own status, Mittag-Leffler advised Poincaré to write a thank-you note, which he himself would deliver to the king. He included a heavy hint that His Royal Highness would be pleased to hear something nice about *Acta Mathematica*'s value to science and about the way in which it was being published. Poincaré did as instructed but, in his haste, forgot to include an envelope for his letter. A few days later, he forwarded the missing item.

News of the prize made headlines all over Europe, and the scientific elite in France were elated. Not only did Poincaré receive the Oscar, but an honorable mention—and another gold medal—went to Paul Appell, also a Frenchman and professor of mechanics at the Sorbonne. What a triumph for the French, after losing the Franco-Prussian War! The dean of the Sorbonne's faculty of science gave an account about the feat to the General Council of Universities, closing with the words that the jury's decision to award the prizes to two Frenchmen was "a new proof of the high esteem in which the most famous among foreign mathematicians hold the work of our French school." The two professors were promptly made knights of the Légion d'Honneur. For good measure, the French government accorded Mittag-Leffler the same honor.

That could have been the end of the story and all participants could have gone on with their lives. But the troubles were only just about to start. A disgruntled scientist, the astronomer Hugo Gyldén, claimed at a meeting of the Swedish Academy of Sciences that he had two years earlier published the same results that Poincaré had presented in his prize essay. Another Stockholm mathematician, Anders Lindstedt, joined the fray with a similar assertion. When King Oscar got wind of the charge, he demanded an explanation. Mittag-Leffler had hoped to have a report on Poincaré's work by Weierstrass in his possession by that time. It would greatly have aided him in a defense of Poincaré's work, but the old and sickly man in Berlin was behind schedule. (On a different level, this may have been just as well, as Mittag-Leffler confided in a letter. The "terrible" Kronecker was just waiting to see the report so he could criticize it.) Somewhat at a loss, Mittag-Leffler asked Poincaré to furnish him with an answer. Poincaré promptly complied with the request by return mail.

Actually Poincaré greatly respected Gyldén's work and maintained that their disagreement was a matter of how convergence of a series of numbers was defined. A series of numbers is an infinite number sequence, such as $1+2+3+\dots$ It is said to converge if its entries add up to a finite sum. For example, $1+1/2+1/4+1/8+\dots=2.0$, hence the series converges. If the sum tends toward infinity as the number of terms increases, the series is said to diverge.

But mathematicians had a different notion of convergence than astronomers. The latter maintained that a number sequence whose entries decrease rapidly in value has—for all practical purposes—a finite sum. For mathematicians this was not enough by a long shot. The terms of the so-called harmonic series, for example, which starts as $1+1/2+1/3+1/4+1/5+\dots$ decrease quickly. Nevertheless, it can be proven that the sum tends toward infinity. It does so slowly: No less than 178 million entries must be added to arrive at the sum of 20, but eventually the sum grows beyond all boundaries. Hence the harmonic series diverges. Thus, even if the terms of the series decrease so rapidly that the computation of a planet's orbit can be computed to a precision of several digits behind the decimal point for very, very long time periods—

we are speaking many millions and billions of years here—convergence of the series must be proven mathematically even as the number of the series' terms tends to infinity.

In their work, Gyldén and Lindstedt had used number series to approximate the orbits of the bodies that move around each other. To establish the stability of a system of three or more bodies—such as our solar system—it is crucial to determine whether the series converge. If they diverge, the system could explode. Gyldén admitted that the series diverge occasionally, but maintained that this happens only for an infinitely small set of parameters. Hence it would be infinitely unlikely that the planets would ever stray from their elliptical orbits. Poincaré, on the other hand, asserted that the set of parameters that lead to explosions, albeit small, could not be neglected. A small disruption—who knows, even a visiting comet—could throw everything off and push Earth out of its regular orbit around the Sun, either into deep space or into a black hole somewhere in the widths of the universe.

In spite of this horror scenario, Mittag-Leffler breathed a sigh of relief. At least he was able to defend Poincaré's work along these lines even if it meant that the stability of the solar system for eternal time was not guaranteed. But this episode was also nothing compared to what was yet to come.

As the rules of the prize stipulated, the winning essay was to be published in an upcoming issue of *Acta Mathematica*. Mittag-Leffler charged his assistant Edvard Phragmén with the preparation of the manuscript for publication. Phragmén was a serious young man and took to his task with religious fervor. He checked every equation, recalculated every computation. In July 1889, half a year after the prize had been awarded, Phragmén came across a point that was unclear to him. He told Mittag-Leffler about it and wrote to Poincaré with a request for clarification.

This time the ball was in Poincaré's court. Not overly worried, he started to follow up on the editor's request. He was already known in mathematical circles for not always abiding by the rigorous rules of the profession. As he had done in his childhood, when his speech was often unintelligible because ideas came to him faster than he could pronounce them, in his mathematical work too he did not always spend sufficient time formulating the details. Much to the consternation of his readers, he sometimes skipped a few steps in a proof.

To satisfy Phragmén, he added explanatory notes that were to be appended at the end of the prize essay. He listed them alphabetically, note A, note B, note C, etc. After not too long, he had already reached note I. But while writing the voluminous notes, he became aware, little by little, of something far worse. Trying to clarify the obscure points, he realized to his utmost horror that his paper contained another error—not a minor slipup, mind you, but a serious defect. He had established that three bodies in a gravitational system can tend toward either equilibrium or periodic orbits or quasi-periodic orbits. But he had overlooked one further possibility: chaotic orbits.

Poincaré was devastated. Here he was, recipient of the Oscar, chevalier of the Légion d'Honneur, darling of the scientific community, and it was all based on an erroneous paper? The significant advancements contained in his paper that remained valid in spite of the errors would forever be forgotten, the new methods that he had invented ignored. All that would remain in the minds of his scientist colleagues and the generations to follow were his mistakes.

Poincaré set about reworking his paper. For months, he worked frantically, but unsuccessfully, on a shortcut around the faulty parts. Gradually the full extent of his errors dawned on him, and eventually there was no more room for doubt. On December 1, 1889, Poincaré took what must have been one of the most painful decisions of his life. He dispatched a telegram to Phragmén asking him to stop the printing process. The same day he also sent a letter to Mittag-Leffler in which he admitted to his great distress. *Je ne vous dissimulerai pas le chagrin que me cause cette découverte,* he wrote (I will not hide from you the sorrow that this discovery causes me).

It was too late. On December 4, three days after Poincaré posted the letter, Mittag-Leffler informed him by return mail that prepublication copies of the faulty paper had already been sent out to a few select personalities. He first tried to make light of the grave situation by congratulating Poincaré on the birth of his daughter Yvonne just a few weeks earlier. But then he got down to business. He deeply regretted the whole matter for Poincaré's sake as well as for his own, he wrote. The noise around the prize had made Mittag-Leffler many enemies who would be more than happy to make a scandal out of the most recent turn of events. Somewhat disingenuously he continued that he considered it "entirely my own fault not to have discovered the weak point in the argument."

He did not really intend to take the blame. To the contrary, his unctuous remark was meant to indicate the exact opposite. Surely nobody could fault him for having missed such a hidden detail. But it was a bit too obvious. He struck the sentence out, writing instead that he did not consider it shameful to have erred when someone of Poincaré's stature had been mistaken. Besides, he wrote, he was sure that Poincaré would someday unravel this difficult question. Maybe the kind remarks were meant to blunt the impact of the bad news; maybe they were meant to encourage Poincaré to continue working on a revision. But when Mittag-Leffler did not feel the need to jolly along one of the most important contributors to the *Acta Mathematica*, he became much more outspoken. In a letter to Hermite, two days later, he was quite blunt: "The error Poincaré made is grave and crucial for the whole paper. There are few pages where he does not utilize the faulty results."

As soon as the letter to Poincaré had gone off, Mittag-Leffler got busy trying to control the damage. Not only his own prestige was at stake. Both Hermite and Weierstrass had missed the error. If word got out, their reputations would irreparably be hurt. Gyldén, one of the recipients of the prepublication copies, would be only too happy to trumpet the failure to the world. Kronecker was lying in wait for just such an occasion. And what about the king? Having been made a laughingstock, his wrath would be implacable. The entire future of the *Acta Mathematica* was at stake. Hoping against hope that Poincaré would somehow be able to rectify things, Mittag-Leffler took pains to cover up all tracks of the error.

He immediately telegraphed to Berlin and Paris to ensure that not one copy of the *Acta Mathematica* would be distributed to subscribers. Then he planned on how to get his hands back on the advance copies that had already been distributed. Only a handful of preprints had been sent out. Obviously, Hermite and Weierstrass had received one each; another one was with the mathematician Camille Jordan in Paris. The editor in chief of the prestigious journal *Mathematische Annalen* had received a copy, as had Sophus Lie in Norway, the astronomer George Hill in New York, two mathematicians in Italy, and a few colleagues in Stockholm. The most problematic recipients of preprints were Hugo Gyldén and Anders Lindstedt.

Mittag-Leffler outlined to Poincaré how he was going to retrieve the advance copies. The operation required much finesse. Apart from Hermite, who already knew what was afoot, nobody would be told anything except that an insignificant error had crept into the printed version of the paper. With some relief Mittag-Leffler confided to Poincaré, "I am very happy that no copy was sent to Kronecker. Concerning the copies that Gyldén and Lindstedt received, I will do my best to get them back without arousing suspicion." He ended his letter with some advice: "I think this whole story must stay between us until your new version of the paper is published. Thus neither science nor you will lose anything through this affair, which will have been treated, I believe, in a quite honorable manner." Not everybody would share Mittag-Leffler's opinion.

Apparently all recipients gave up their copies without any fuss and nobody was the wiser. Except Weierstrass, that is. He had been kept in the dark about the embarrassing affair. Mittag-Leffler's sincere concern about the old man's health may have kept him from telling him the news. But rumors reached Berlin. Hugo Gyldén had come for a visit and was all too happy to tell everything he suspected, to anybody who would listen. And Kronecker was quite willing to listen. So when Weierstrass was asked by his colleagues about a possible error in Poincaré's prize essay, he found himself in an extremely uncomfortable position. Nobody would believe his assurance that he knew nothing about it. Not only did Weierstrass feel frustrated that a hidden error existed somewhere that he himself had not spotted when he was supposed to act as a judge, he also had the distinct feeling that the wool was being pulled over his eyes. In the telegram that asked him to return the advance copy, mention was made only of a small error, the correction of which would entail a slight delay in the publication schedule. Now he realized that there was far more to the matter than had been revealed to him.

Understandably upset, he wrote an indignant letter to Mittag-Leffler. His upright Germanic spirit had already bristled with misgivings when he heard that post-award changes would be included in the published version in the paper. Now it turned out, not only were explanatory notes to be added, but a whole new version was being written to correct an error. Mittag-Leffler tried to calm him down. The error was by far not as serious as Gyldén tried to make out, he wrote. In fact, the astronomer from Stockholm probably found it convenient to exaggerate its magnitude to further his own interests. Moreover, by rewriting his paper, Poincaré would make it more accessible and it would gain a lot. Once it appeared in print, only people like Gyldén, who had taken a liking to diverging series, or like Kronecker, who would not accept anything if it had not been invented by himself, would not be among its admirers.

Weierstrass was not placated. Unwilling to treat the matter as lightly as Mittag-Leffler, Hermite, and Poincaré were about to do, he sent lengthy complaint letters to Stockholm. In Germany, he wrote, it was out of the question to allow changes to a prize essay once a decision had been taken. After all, later readers should be able to form their own opinion based on the exact same paper that had been presented to the jury. Of course, it pained him that he had not noticed the error himself, he continued, and then delved into a little damage control of his own. He did not want to mention that he had been ill at the time, he wrote, but then went ahead and mentioned it. More to the point, nobody could demand from a juror that he guarantee the truth of every statement in such an extensive manuscript, especially when many computations were not actually performed but only hinted at. The latter was a thinly disguised swipe at Poincaré's irritating habit of omitting whole sections of a proof because they seemed clear to him. Furthermore, Weierstrass wrote, nobody should fault a reader because he'd relied on the mathematical competence of the author. Besides, the difficult result, which had now turned out to be incorrect, was so appealing that Weierstrass had let himself be led astray. The most annoying matter was, however, that the report to the king would now have to be withdrawn. And so it went, on and on and on and on.

Mittag-Leffler had little time for the cranky old man; the more important cleanup operation was in full swing. So far, everybody had accepted his explanation that a small error had been overlooked in the galley proofs and that Poincaré insisted on correcting it. But Mittag-Leffler was concerned about Gyldén's copy. Finally, he decided to pick it up in person. He was successful, but then gave free rein to his frustration in a letter to Hermite. The error was serious, he wrote, and Poincaré needed to do substantial work, far beyond a simple revision. But it would serve him a good lesson. Maybe he would finally abandon his infuriating habit of announcing results whose proofs he himself knew only imperfectly. Maybe the episode, as soon as it became public knowledge, would lead others to reexamine more critically his previous works. Certainly Poincaré was a genius, but this time he had inflicted too much on everybody. He was still young—Poincaré was then thirty-two years old—and he would change. Mathematical science would be the better for it, Mittag-Leffler concluded.

The covert operation had succeeded beyond Mittag-Leffler's expectations. All outstanding preprints containing the faulty essay had been returned, and the whole print run of volume 13 of the *Acta Mathematica* was pulped. The only surviving copies of the original printing are tucked away at the Mittag-Leffler Institute in Djursholm, outside Stockholm.

Now it was up to Poincaré. Work had to be done that went far beyond a simple revision of the essay. Not knowing whether he would succeed, Poincaré set out to invent the new tools and methods that were required. After several months of intensive work, he was ready. The new manuscript was truly pathbreaking. It provided the first traces of chaos theory, which would become so popular only a century later. The possibility of chaotic movement of bodies was just what he had overlooked in the erroneous first version of the essay. It turned into a fateful lead indeed. For example, it is the basis for the so-called butterfly effect, which states that small causes can lead to great consequences: A butterfly flapping its wings in Texas could provoke a thunderstorm in Australia.

The new manuscript also includes Poincaré's famous "recurrence theorem," which asserts that an energy-conserving system of bodies, left to itself, will eventually return to its initial position (or to an arbitrarily close position). This implies—to mankind's immense relief—that even if the planets of our solar system were to travel very, very far away from one another, they would eventually return to their initial positions, though it could take a very, very long time. In January 1890, Poincaré sent the new version of the paper to Mittag-Leffler. It was a hundred pages longer than the initial essay.

At Mittag-Leffler's urging, Poincaré had added only a caustic reference to the error: In the introduction he thanked Edvard Phragmén, who had called his attention to a *point délicat* (a delicate point), which had allowed him to discover and rectify an important error. That was it. Nobody who read the paper would notice anything amiss, and those in the know kept mum about the real extent of the near disaster. As requested, Poincaré paid for the second printing of volume 13. On June 1, 1890, he sent a money order for 3,585 crowns and 65 ore to Mittag-Leffler. That was 1,085 crowns and 65 ore more than the prize money he had received, but Poincaré was only too happy to oblige.

On the whole, Mittag-Leffler's scheme had worked out just fine. When volume 13 of the *Acta Mathematica* was printed in April 1890, nobody remembered the rumors of an erroneous early version of Poincaré's paper. The masterly essay alone remained. Only many years later, when historians of science searched through the archives of the Mittag-Leffler Institute and compared the original version with the published paper, was the true magnitude of the differences discovered.

For Mittag-Leffler it was time to take a well-deserved rest. The initial plan had been to announce a new prize essay every four years, but unsurprisingly, after this fiasco, no more mathematical Oscars were ever awarded. Phragmén, the assistant who had set the ball rolling, was appointed professor at Stockholm University with the active help of Mittag-Leffler and Poincaré. He later did work on insurance mathematics, became director of an insurance company, and served as president of the Swedish Society of Actuaries for ten years. Poincaré went on to hammer out more details about the stability of the solar system during the following years. The three volumes of his twelve-hundred-page opus *Les méthodes nouvelles de la mécanique céleste* (New Methods of Celestial Mechanics) appeared successively in 1892, 1893, and 1899. In 1967 an English translation of the landmark work was published by NASA, and the American Institute of Physics issued a reprint a century after the appearance of the first volume in Paris.

In the excitement about the prize and the associated hullabaloo, let us not forget our poor solar system. Is there anything to worry about? Both versions of Poincaré's prize essay left the ultimate question of whether the n-body system could explode open. One thing Poincaré did discover was the ominous butterfly effect, that minuscule disturbances in a system could have enormous consequences. Hence, it was just conceivable that the perturbations caused by an object as small as a spacecraft could tear planets asunder. The question of the solar system's eternal stability was still open.

In 1912, Karl Sundman, a Scandinavian mathematician, found an infinite series that describes the orbit of the bodies in a system and does converge—albeit infinitely slowly. In 1954, the Russian Andrei Nikolaevich Kolmogorov gave a plenary lecture about the n-body problem at the congress of the International Mathematical Union in Amsterdam. The subject was the thorny question of what happens to periodic orbits of bodies when small perturbations disturb their course. His answer was that many perturbed orbits could become quasi-periodic but that they would not explode. Hence, apart from small wobbles, the solar system will remain stable. So, no need to worry, we can sleep quietly.

Can we really? Well, not quite. At the Courant Institute in New York, the German Fulbright scholar Jürgen Moser was asked by an editor of *Mathematical Reviews* to write a report on Kolmogorov's paper. And just like Phragmén sixty-five years earlier, Moser became suspicious when he stumbled over a passage that was unclear to him. As Moser studied it more closely, it seemed to him that Kolmogorov's central thesis was not completely proven. He set out to fill in the gap. For six long years Moser worked on the problem before he could definitely close the missing gap in Kolmogorov's proof. He solved a problem of "small divisors" that had already plagued Poincaré by showing that the numerators in the series diminished faster than the divisors and hence the series converges. Actually, a controversy exists to this day about whether Moser's contribution was really needed. Some mathematicians maintain that Kolmogorov's proof was not lacking to start with.

At the same time that Moser was looking for the missing piece of the puzzle, one of Kolmogorov's students, Vladimir Igorevich Arnold, attacked the n-body problem from a different angle. While Moser considered perturbations to the orbits that were sufficiently smooth, Arnold investigated perturbations that could be expressed as number series. He arrived at the same conclusion as Moser. Thus our solar system's stability could finally be ascertained: During our, our children's, and our grandchildren's lifetimes, nobody has to worry. At least for the foreseeable future, planets will not deviate from their orbits. In honor of the three mathematicians—Kolmogorov, Arnold, and Moser—the new theory was named after their initials: KAM theory. In spite of their comforting results, a queasy feeling remains nonetheless: The stability of our solar system, which consists of nine planets, not just three, is far from guaranteed by science.

Armed with the Oscar, Poincaré stepped back from the limelight to devote himself again to teaching and research. His output was prodigious. In just three and a half decades he published no less than five hundred papers, memoirs, and books. (By all accounts, Poincaré was a prolific writer…and a good son. He wrote more than three hundred letters to his mother while a student in Paris.) But his life's work was vast not only in number but also in the areas it spanned. Poincaré was one of the world's last two mathematicians who had a full understanding of all existing branches of the field. The other one was David Hilbert in Göttingen. After the demise of these two giants, in the first half of the twentieth century, the subject split into so many disciplines and subdisciplines that no single mathematician could ever again hope to grasp more than just the subject matter in his or her immediate area of specialization.

At the turn of the nineteenth to the twentieth century, it was still possible to be a universalist, provided you were a genius of Poincaré's stature. His seminal writings encompassed many branches of mathematics, physics, and the philosophy of science. Ideas came to him intuitively. Solutions to the problems he was thinking about sometimes appeared out of the blue, during a sleepless night after having drunk too much black coffee, while pacing corridors, or stepping on a bus. As soon as he had something approximately worked out in his head, he put it to paper. Consequently, his writing style often lacked rigor. His papers were often chock-full of hints and ideas that were not always sufficiently proven. Whenever he had completed a paper, he immediately regretted its content and style.

In mathematics his contributions are too numerous to list. Among the work that had the greatest impact on mathematics are his six papers on analysis situs, the subject that is today known as topology. Attempting a geometric approach to the description and classification of the solutions to algebraic equations, Poincaré was led away from the manipulation of numbers and formulas and toward the visualization of curves and flows. The subdiscipline he thus pioneered was to become algebraic topology. It will concern us throughout the remainder of this book.

In physics he dealt with celestial and fluid mechanics, optics, electricity, capillarity, elasticity, thermodynamics, theories of light and of electromagnetic waves, quantum theory, and the special theory of relativity. Poincaré's work on the latter subject is of particular interest. At the same time that the patent clerk third class Albert Einstein was working on the special theory of relativity in the Swiss town of Bern, Poincaré came close to discovering the theory himself. As might be expected, there were, and still are, some nationalistically inspired attempts to reclaim the theory for France. However, the consensus is that Poincaré fell short. Even though he had the necessary mathematical tools at his disposal, he was not audacious enough to make use of them in the revolutionary way that Einstein did.

Poincaré's teaching style was not universally appreciated. As usual, his mind worked too fast, and his lectures were sometimes ill-prepared and lacked polish. It did not help that he apparently got bored with a course after he had taught it once. Since he never offered the same subject twice and never looked back, he never had occasion to straighten out the frayed edges. It was left to students to edit his lecture notes and have them printed as textbooks, which then turned out to be very successful. Amazingly, Poincaré had only a single doctoral student, Louis Jean-Baptiste Bachelier, whose thesis, *Théorie de la spéculation* (Theory of Speculation), is considered one of the seminal works of modern financial theory.

Poincaré died on July 17, 1912. A first sign that all was not well with his health had already appeared at the International Congress of Mathematicians in Rome in 1908. Just before he was to give a talk titled "The Future of Mathematics," he was suddenly struck by prostate problems. The talk was delivered by a colleague while Poincaré was treated in the local hospital. Mrs. Poincaré hurried to Rome to bring her husband back home. Back in Paris, Poincaré soon resumed his old schedule and everything was normal for another four years. But on July 9, 1912, he had to undergo surgery. He seemed to recover well, but eight days later, an embolism apparently caused his sudden death. He was only fifty-eight years old.

During his lifetime, Poincaré received many honors; the Oscar was just the first and the knighthood in the Légion d'Honneur the second. In 1887 he was elected to the Académie des Sciences, becoming its president in 1906. Two years later, in 1908, Poincaré became a member of the Académie Française, the gathering place of the most famous French writers and poets. A small question mark may be attached to his membership in that august society. While there is no doubt about Poincaré's achievements in science, his status as a literary giant comparable to the likes of Racine and Voltaire is dubious. His three more popular philosophical works *La science et l'hypothèse* (Science and Hypothesis, 1901), *La valeur de la science* (The Value of Science, 1905), and *Science et méthode* (Science and Method, 1908), while of great interest—they were said to have even been read by secretaries and other womenfolk—may have served primarily as a pretext for his election.

Rumors have it that there was another reason for his election to the hallowed halls. Raymond Poincaré, Henri's cousin and at various times president or prime minister of the French Republic, was a candidate for entry to the Académie. Some of his political opponents wanted to block him. Thinking that the academy's bylaws did not allow two members of the same family to be members, they voted to elect Henri. They were mistaken. No such bylaws existed, and Raymond was duly elected a year after his cousin. Another rumor has it that the academy's famed dictionary commission needed someone to sort out all the new words in physics and mathematics. Speaking of family, Henri's brother-in-law, the philosopher Émile Boutroux, was also elected to the Académie Française, albeit three months after Henri died.

Apart from the numerous honors, medals, and prizes that Poincaré was awarded during his lifetime, there are many landmarks in France and even in outer space that remind us of him. The university in his hometown of Nancy, for example, is named after him, and so is one of the town's best-known *lycées*. In Paris, the Rue Henri Poincaré in the fashionable twentieth arrondissement honors the great mathematician—admittedly, Cousin Raymond has a tree-lined, four-times-as-long avenue named after him, but he was president, after all—as does the Crater Poincaré in a not-yet-fashionable part of the moon. (No crater for the president cousin, though.) While we're at it, there is also the Avenue du Recteur Poincaré, in memory of Henri's cousin Lucien, an important university administrator, and in the thirteenth arrondissement an avenue carries the name of his brother-in-law, Émile Boutroux.

Chapter 5

Geometry Without Euclid

Three mathematicians are shown a cube and asked to describe what they see. The first, a geometer, says, "I see a cube." The second is a graph theorist. She ventures, "I see eight points, connected by twelve edges." The third, a topologist, declares, "I see a sphere." The joke encapsulates the worldviews of mathematicians belonging to different disciplines. Everybody sees what he or she wants to see and is blind to much else. Topologists are blind to angles—or the lack thereof—distances, and the exact shapes of the objects of their interest.

As every schoolchild knows, Euclid's geometry requires them to measure, or at least compare, angles and lengths of geometrical objects. In the eighteenth century a revolution took place that freed geometry from the constraints of measurement. The new discipline, which would eventually be called topology, describes the character of geometric objects without taking recourse to measurements. The revolution started with the work of the Swiss mathematician Leonhard Euler.

Euler, one of history's most important mathematicians, was born in 1707 in the Swiss city of Basel. He was the pupil of another remarkable Swiss mathematician, Johann Bernoulli. In 1726, he was sent by his teacher to the academy of St. Petersburg in Russia to keep company with yet another remarkable Swiss mathematician, his son Daniel Bernoulli. The Bernoulli family produced a few additional remarkable men of science. In fact, the position in St. Petersburg had become available because the previous incumbent, Daniel's brother Nikolaus Bernoulli, had just died.

Euler was prolific, both in offspring and in intellectual output. He fathered thirteen children, albeit with two wives, and wrote more than eight hundred books and papers in all areas of mathematics. This is all the more astonishing—the part about the papers, that is, not the children—since for a large part of his life he was blind. His power of concentration must have been nothing less than astounding, keeping in mind that he did much of his work without eyesight while screaming kids were scampering around. Late in life he claimed that he had done some of his best work with a baby in his arms and other children playing at his feet.

While at the Academy of Sciences in St. Petersburg, Euler received an intriguing letter from the picturesque town of Königsberg in Prussia (Kaliningrad in today's Russia). With branches of the river Pregel cutting through it, the city consisted of four separated quarters, connected by seven bridges. The sometime mayor of the nearby town of Danzig, Carl Leonhard Gottlieb Ehler, wanted to plan a special walking tour through the town in such a way that tourists would cross all the bridges. He was not concerned about the path's length—after all, if tourists were worried about strolls being too long, they might as well stay home—but he did not want to bore them. Therefore, the strollers should cross every bridge exactly once. But try as he might, Ehler could not find a suitable path, and after a while he stopped trying. He sought help and somehow had the feeling that if anybody could find the appropriate path, it would be the famous mathematician in St. Petersburg. So he sent a letter to Euler: "You would render me a most valuable service if you could send us the solution to the problem of the seven Königsberg bridges, together with a proof."

The great mathematician balked at first. Why should he waste his time with trivia? This was no question for a mathematician but rather for promoters of tourism, town planners, or tour guides. In a somewhat

indignant letter he answered Ehler that the problem "bears little relationship to mathematics and I do not understand why [solutions] are
expected from mathematicians rather than from anybody else, since
they are based solely on common sense." In a rhetorical aside that
smacked just a wee bit of false modesty, he continued, "Why is it that
questions that have very little relationship to mathematics are solved
faster by mathematicians than by other people?" He left the question
unanswered and the we-are-just-smarter implication was left to hang.

Euler may have had a point in his initial assessment that this was no
question for a mathematician. Indeed, no tools were available in all of
mathematics to tackle this kind of conundrum. Traditional geometry
dealt with lengths and angles, which were of no use in the case of the
Königsberg bridges. True, the German philosopher Gottfried Wilhelm
Leibniz (1646–1716) and his pupil Christian Wolff (1679–1754) had
suggested a new mathematical discipline, which they called analysis situs, that could possibly shed some light on such problems, but hardly
anybody had even heard about it.

Obviously, the mere lack of tools could not stop the great Euler, once
he got interested. Apparently, Ehler had tickled a nerve. Only four days
after receiving the letter, Euler wrote to the Italian mathematician
Giovanni Marinoni, court astronomer in Vienna. After describing the
problem, he wrote:

> I was informed that hitherto no-one had either demon-
> strated the possibility of finding a route, or shown that it is
> impossible. This question is quite banal, but it seemed to me
> worthy of attention in that neither geometry, nor algebra,
> nor even the art of counting was sufficient to solve it. In view
> of this, I wonder whether it belongs to the geometry of posi-
> tion, which Leibniz had once so much longed for.

In this, Euler was not quite forthright. The thought that Leibniz's
geometry of position was the appropriate tool had not occurred to Euler,

but rather to Ehler, who had mentioned in his letter that the Königsberg problem "would prove to be an outstanding example of the calculus of position, worthy of your great genius." To this, Euler had replied, "I am ignorant as to what this new discipline involves, and as to which types of problem Leibniz and Wolff expected to see expressed in this way." Be that as it may, Euler was quick to catch up, even if it was not his idea.

He soon figured out that the mayor was right on two counts. First, as Ehler had suspected, the desired foot walk simply did not exist in Königsberg. Second, it really was Leibniz's geometry of situation that provided the tools to investigate the question. Actually Euler did more than just provide a yes/no answer to a vexing problem. He found the exact conditions under which a town with islands and bridges can be perambulated in the prescribed manner, and under which circumstances it cannot. "After some deliberation," he informed Marinoni, "I obtained a simple, yet completely established rule with whose help one can immediately decide for all examples of this kind, with any number of bridges in any arrangement, whether such a round trip is possible, or not." What he found, and proved, was that the existence of the desired path depends on the number of areas to which an odd number of bridges lead: A path exists if the number of bridges is odd for no town area or for two areas. No path exists if there are either one or more than two areas to which an odd number of bridges lead.

Euler's paper *Solutio problematis ad geometriam situs pertinentis* (Solution to a problem concerning the geometry of position), published by the Imperial Academy of Sciences of St. Petersburg in 1736, can be said to have heralded the new mathematical disciplines of graph theory and topology. Until then, walkers were concerned—and understandably so—with taking the shortest paths, builders with constructing right angles, artists with drawing round shapes. Here the lengths of paths were of no interest, and neither were the exact shapes of the town's four quarters. All that mattered was how the different parts of the town were connected by bridges.

Fourteen years later, in 1750, Euler struck again. Of course, Euler did many things in between, but here we are interested only in work that relates to topology. In a letter to his colleague Christian Goldbach (1690–1764) he reported a remarkable fact. While investigating polyhedra, he had discovered that the number of vertices, edges, and faces of straight-edged solids always satisfies a simple formula: Whatever their shape, when deducting the edges from the vertices and adding the faces, one always gets 2. A cube, for example, has 8 vertices (corners), 12 edges, and 6 faces. A pyramid with a square base has 5 vertices, 8 edges, and 5 faces. A pyramid with a triangular base has 4 corners, 6 edges, and 4 faces. An icosahedron has 12 vertices, 30 edges, and 20 faces. All these solids, as well as infinitely many others, give 2 as the answer to Euler's formula. Try it with a cube with one corner whacked off, a cube with all corners whacked off, a pyramid whose base is not square or triangular but twenty-cornered. You won't be disappointed.

At first, Euler did not know why this was so, and he published a paper in 1752 in which he simply stated the finding as a fact. This was quickly followed by another paper in the same year in which Euler proved his formula by dissecting the solids into slices.

But there were problems with the formula. First of all, Euler had not been the one to discover it. The French mathematician and philosopher René Descartes had already discovered an equivalent theorem 130 years earlier. But Descartes' original manuscript on the subject was lost, and his discovery came to light only when a copy was discovered among the papers that Leibniz left after his death.

Second, and more important, the formula is not always valid. The first to notice that something was amiss was the French mathematician Adrien-Marie Legendre in 1794. The notion of a convex body must be explained to see what was wrong. A body is convex if a straight line that connects any two points of the body lies wholly inside the body. Take a cube, for example, and choose any two points—say, opposite corners. The line connecting the corners lies wholly inside the cube; hence the solid cube is convex. Now drill a hole through the cube. Some lines that start and end in the cube traverse the hole. This means that part of the line lies outside the body, and the cube with the hole is not convex.

As it turned out, Euler had considered only convex solids, overlooking that bodies could have indentations, holes, or cavities. And for some—not all—such nonconvex solids, Euler's formula was incorrect. Take, for example, a cube that contains hidden within it an empty cube-like cavity, like a sealed burial chamber. It is nonconvex, since some straight lines passing through the cube must pass through the void in the cavity. The number of vertices is 16, the number of edges is 24, and the number of faces is 12. Uh-oh, that gives us 4, not 2. Something is definitely wrong. Apparently, Euler was somehow off the mark.

It took another Swiss, Simon Antoine Jean l'Huilier (1750–1840) from Geneva, to provide the missing links. As a pupil, l'Huilier excelled in mathematics without, however, reaching the level of his famous compatriots. When he was a young man, a wealthy relative offered to make him the heir to a large fortune if he would enter a career in the church. But l'Huilier, who had already taken a liking to mathematics, would not hear of it. He studied mathematics in Geneva under one of Euler's students, then went on to become private tutor to the children of a wealthy family. His break came when he entered a call for writers by the Military Academy in Warsaw, seeking textbook authors for the Polish school system. L'Huilier won and was commissioned to write a textbook on mathematics.

One of the organizers of the competition was Prince Adam Kazimierz Czartoryski. The British-educated prince was being prepared for the throne of Poland, but refused to take office. He later became the first minister of education in a European country. Czartoryski was so impressed by l'Huilier's proposal that he invited him to his palace in the town of Pulawy, in eastern Poland, and appointed him tutor to his son.

L'Huilier stayed at this post for eleven years. It must have been a cushy job because his teaching duties left him enough spare time to do research in mathematics and to write mathematical papers. He also entered a prize contest that the Berlin Academy of Sciences announced in 1784 on differential calculus. L'Huilier won first prize with an essay on the notion of infinity and mathematical limits. After an interlude at the University of Tübingen he was appointed to the chair of mathematics in Geneva in 1795, where he stayed until his retirement, at age seventy-three, in 1823. As a resident of Geneva, l'Huilier did not neglect his civic duties. He was involved in local politics and even presided over the city's legislative council for a year. He also served as rector of the Academy of Geneva.

As a mathematician, l'Huilier was dedicated and proficient, but far from the stature of an Euler or a Bernoulli. This did not keep him from filling in the gap the great Euler had left open. L'Huilier reexamined the formula and found a new one that takes into account hidden cavities. He showed that the 2 on the right-hand side of Euler's formula must be replaced with 2 plus twice the number of cavities. This new right-hand side is nowadays called the body's "Euler characteristic."

Let us consider again the cube that contains a small, empty cube inside it. Counting the items one finds 16 vertices, 24 edges, 12 faces. Hence the Euler characteristic computes to 4, which is just 2 plus twice the cavity. Next try a pyramid with a pentagonal base and a triangular pyramid inside it. This solid has 10 vertices, 16 edges, 10 faces. The Euler characteristic again equals 4. Try any shape you like and give the inside cavity any profile you like. The Euler characteristic always remains 4.

This is the most intriguing feature about the formula: Cubes can be replaced by pyramids, icosahedra can morph into dodecahedra, the solids can be right-angled or skewed, but the Euler characteristic never changes. In the absence of cavities it is always 2. If a cavity hides inside the solid, the Euler characteristic is 4, no matter what shape the body and the cavity. If the body contains two cavities, the Euler characteristic is 6. Again it does not matter whether the body is a cube or an icosahedron and the cavities are two octahedra, or an octahedron and a pyramid, or a cube and a tetrahedron, or anything else. If, in addition to hidden cavities, the body contains bulges, indentations, and tunnels, the equation has to be further adapted, but everything we are about to say still holds.

Apparently, the formula tries to tell us something. All solids—no matter what their shapes—are in some way equivalent to one another, but are different from bodies with a cavity. The latter, in turn, are equivalent to one another, whatever the bodies' shapes and the cavities' profiles, but are different from bodies with two cavities. And so on. Could l'Huilier's formula be used to classify bodies?

Indeed it can. With the help of the formula all kinds of bodies can be sorted into well-defined categories. Solids can be lengthened or shortened, squashed and squished, but the Euler characteristic does not waver. Only when cavities, tunnels, dents, or bulges appear does its value change. Then these cavities and tunnels and dents and bulges can be turned and twisted and warped and skewed without affecting the Euler characteristic. The Euler characteristic provides what would become known as an invariant: Its numerical value does not change, even when the body is transformed into seemingly quite different shapes. In some way this is like the Königsberg problem. The town's islands could be small or large, round or square, oblong or prolate, without affecting the existence, or nonexistence, of an appropriate path.

Once this kind of geometry began to establish itself as a serious field for mathematicians, a name had to be found for the new theory. *Calculus of position* and *analysis situs* just weren't very catchy. After all, when asked what one does for a living, it just does not sound right to say that one is a "calculator of position" or a "calculus of positionist." It fell to Johann Benedict Listing from the University of Göttingen to coin a name.

Listing was born in 1808 in the German city Frankfurt am Main. The only child of a poor brush maker and a peasant woman, he distinguished himself as a young boy by his talent for the arts. From the age of thirteen, he helped his parents augment their meager income by doing drawings and calligraphy for the public. Despite his modest upbringing he attended good schools and received a sound education in modern and classical languages, mathematics, and the sciences. Such were his achievements at school that he was awarded a scholarship that enabled him to continue his studies at university. Since the stipend was granted only for the study of arts, not science or mathematics, Listing chose architecture as a compromise.

He entered the University of Göttingen in 1830 and took up a broad range of subjects. Apart from the required courses in architecture and the chosen courses in mathematics, he attended lectures in astronomy, anatomy, physiology, botany, mineralogy, geology, and chemistry. The bright and hardworking student soon came to the attention of Carl Friedrich Gauss (1777–1855), the undisputed prince of mathematics of the nineteenth century. He was one of the few mathematicians throughout history to be equal in importance to Euler. Listing did his doctorate under Gauss's supervision, submitting a dissertation on geometry. In July 1834, less than a month after completing his thesis, he set out on an expedition with his friend the geologist Wolfgang Sartorius von Waltershausen, to study the volcano Mt. Etna in Sicily. After procuring the necessary equipment, they set out southward. They passed through Karlsruhe, Stuttgart, Munich, Salzburg, Innsbruck, Verona, Milan, Venice, Rome, and Naples. Not surprisingly, it took them a year to reach Sicily. Sartorius could finally perform the surveys and measurements that would form the basis for his magnum opus on Mt. Etna.

During the trip Sartorius fell seriously ill, and the local doctors thought he would die. But Listing nursed him back to recovery. Then Listing fell ill, and it took him a month to recover. Now they were ready to return home, but by that time everybody else had fallen ill: A serious outbreak of cholera made the Italian mainland inaccessible. So the two travelers set out for home on a circuitous route. First they caught a Danish boat that was headed for Rio de Janeiro with a cargo of Sicilian wine. Upon reaching Gibraltar, they disembarked and took another boat to Lisbon. From there they sailed on to Liverpool. This unplanned detour called for another layover, and they stayed in Britain for a couple of weeks. Only in the fall of 1837 did they return to Hanover.

While abroad, Listing had received by mail an offer to teach applied mathematics, machine design, and engineering drawing at a vocational school back in Hanover. At first, he did not feel like committing himself, but upon his return from his adventurous journey, a laid-back job was just what Listing needed. He accepted the offer, took a liking to his new occupation, and looked forward to a comfortable career as a teacher.

He could have remained in this post for the rest of his life, except for a seemingly unrelated event across the Channel, in England. In the same year that Listing started his job at the vocational school, King William IV died, and because he had no heirs, his niece Victoria became queen of England. For over a century, the king of England had automatically been ruler of the Kingdom of Hanover, to which the town of Göttingen belonged. However, Hanover's ruler had to be a man. Thus the queen's uncle, the Duke of Cumberland, became king of Hanover instead. For the professors of the University of Göttingen this spelled trouble. The new ruler, King Ernst-August, demanded a pledge of allegiance from his kingdom's civil servants, among them the university professors. While the conservative Gauss did not mind pledging, his close associate and friend the physics professor Wilhelm Weber, as well as six others, refused. The "Göttingen Seven" were promptly fired.

For a long time, Weber's university post remained vacant. But after two years, it was no longer possible to keep physics absent from the curriculum of a respected university; the post needed to be filled. Weber was still living in Göttingen as a private citizen—all the while working with Gauss on physical experiments—but the king would not allow him back and a replacement had to be found. To add insult to injury, it was Gauss, Weber's good friend, who was asked to suggest a substitute. He put forth three names. Last on his list was his former student Johann Listing. Since the two first-ranked candidates refused the university's offer, only Listing, who had never published anything whatsoever in physics, remained. The cabinet secretary of Hanover who was in charge of the affairs of the university summoned Listing to his office and offered him Weber's post. Listing was stunned, but not so stunned that he didn't quickly accept. Thus Queen Victoria's ascension to the throne secured an academic position for Listing. It enabled him to spend his productive life doing research instead of instructing high school students.

As a tenured professor, Listing could choose any subject he liked for his research. He picked a topic whose importance had repeatedly been pointed out to him by his mentor: the calculus of position. Gauss himself never published anything on this theme, but for his student it was just the right subject. In 1847, his *Vorstudien zur Topologie* (Preliminary Studies on Topology) appeared. The book was by no means a pathbreaking work and did not pretend to be more than its title indicated: a preliminary study for a new scientific field. Its most distinguishing feature was that this was the first time the new field's title appeared in print. "By topology we mean the study of the features of objects," Listing wrote, stressing that these features were to be studied "without regard to matters of measure or quantity." The name stuck, namely because *topological* or *topologist* sounded better than anything that could be done with *analysis situs*.

In 1848 a revolution swept Hanover, and King Ernst-August was forced to agree to some measures of liberalization. A year later, Weber was allowed to return to his teaching position. But Listing had served the university well for nearly ten years and it would have been unfair to remove him from his post simply because his predecessor was to be given back his old job. The university administration reached a Solomonic decision: Weber became incumbent of the newly established chair of experimental physics, while Listing kept his professorship, which was renamed the chair for mathematical physics.

For all the good fortune, in his personal life Listing was not happy. He had taken a disagreeable woman as wife. Pauline, née Elvers, was a beguiling woman, fifteen years younger than her husband. The daughter of a lawyer from a nearby town, she met Listing when visiting Göttingen with her mother. A nine-month courtship followed and the couple got married in September 1846.

Problems started within three weeks after the wedding when Pauline ran out of the household allowance that should have lasted a full month. In addition to being a spendthrift, she was also incapable of handling the domestic staff. So severely did she abuse her servants that she was forced to appear several times before the town's magistrate. Spats with landlords led to many relocations. Most of the time, the family—Johann and Pauline had two daughters—lived way beyond its means. To maintain the standard of living that his wife wished for, Listing had to borrow money, sometimes from usurers, and was not always able to repay his debts. More court appearances followed, and the family was saved from bankruptcy only with the help of Sartorius von Waltershausen. The old friend, recalling with gratitude how Listing had nursed him through his illness thirty years previously, arranged for a moratorium by the authorities of Hanover. The financial problems diminished somewhat when Listing received Gauss's rent-free apartment at the Göttingen observatory after the latter's death in 1855. His neighbor at the observatory, with whom he shared a terrace, was none other than Gauss's most celebrated former student, Bernhard Riemann.

Listing's contributions to science were many, apart from mathematics. They included studies in geology, geodesics, optics, meteorology, magnetism, spectroscopy. By studying the optical properties of the eye, he became one of the pioneers of modern ophthalmology. He even invented a method to determine the amount of sugar in the urine of diabetics. In spite of his being gregarious, witty, and good-natured, Listing was ostracized by his colleagues due to his wife's unbearable behavior. Because of that, his scientific achievements probably received less recognition than they deserved. Nevertheless, he remained active in a host of areas throughout his life, teaching, writing, and directing dissertations. Listing died of a stroke on Christmas Eve 1882.

A much more important contribution to the young field of topology than his *Vorstudien* was a paper that he published in 1862 in the *Abhand-lungen der Königlichen Gesellschaft der Wissenschaften zu Göttingen* (Proceedings of the Royal Society of Sciences in Göttingen) in 1862. The paper was entitled *"Der Census räumlicher Complexe oder Verallge-meinerungen des Eulerschen Satzes von den Polyedern"* (Inventory of Spatial Objects or Generalization of Euler's Theorem on Polyhedra). The volu-minous paper comprised no less than eighty-six pages and sixty-four il-lustrations. Since Listing was dealing with a wholly new field, he had to invent a host of words. To make his paper more readable, he appended a glossary, in which the definitions of thirty-one new terms were given.

Listing's paper did much more than just take stock of spatial ob-jects. For example, it investigated the connectivity of surfaces in three-dimensional space. This is, of course, just the point of view that Euler had pioneered in his solution to the problem of Königsberg. Lengths and angles played no role, neither in the way the problem was posed, nor in the solution. All that mattered was how the different parts of the town were connected by bridges. Now Listing generalized this viewpoint.

The *Census* also anticipated many future developments in topology—for example, the so-called Alexandroff compactification, or the differ-ence between homeomorphism and homotopy. Recently, a historian of mathematics speculated that some topological discoveries could have been brought forward by many decades had mathematicians only both-ered to contemplate the illustrations in the *Census* more thoroughly. But the paper, hidden away among works on botany, medicine, philology, history, and various other treatises, was not easily accessible, and this is why one of Listing's most famous discoveries is nowadays not even re-membered as having been his. In Figure 1 of the *Census*, Listing depicted a two-dimensional object floating in three-dimensional space that only has one side. It is, of course, the Möbius strip. But even though Listing included it in his *Census*, he did not recognize its amazing properties. All he did was mention in a footnote that the strip had quite different properties than other two-sided objects.

It was left to August Ferdinand Möbius (1790–1868) to make the

strip into a household term. Educated at home until age thirteen, Möbius—a descendant of Martin Luther on his mother's side—showed
great interest in mathematics when he entered the formal school system.
Nevertheless, upon enrolling in the University of Leipzig, he began to
study law. But his interests soon shifted back to mathematics, as well as
to physics and astronomy. Möbius also went to Göttingen to study astronomy under Gauss. In 1815, when he was twenty-five years old and
had just completed his doctoral dissertation, he was to be drafted into
the Prussian army. He managed to avoid this unsavory experience and
instead completed his *Habilitation*—the thesis that entitles scientists to
teach at German universities.

In spite of Möbius's penchant for mathematics, his academic appointments were in astronomy. Unfortunately, he was not a good lecturer. Even though his income derived from the fees that students paid,
he announced that his classes would be free of charge. Otherwise, he
feared, nobody would have come to listen to his lectures.

The professor remained a bachelor and lived with his mother until
she died. Only after her death did Möbius, by then thirty years old, take
a wife. The poor woman became blind soon thereafter but nevertheless
raised the couple's three children, a daughter and two sons, with love
and care. The two boys would later become noted literary scholars.

In 1844 Möbius was appointed professor of astronomy at the University of Leipzig, and four years later he became director of the Observatory of Leipzig. Möbius did serious work in astronomy, but his most
important contributions were to mathematics. He was methodical and
would not be rushed to publish. Unfortunately, his contributions were
not always original, since he did not bother to keep abreast of the current literature. Thus he had to find out, to his sorrow, that some of his
work had been anticipated by others. His most important offerings were
the design of simpler and more efficient methods for existing subjects.
His manner of writing fitted the German-scholar stereotype and was
described by a biographer as follows: "He worked without hurrying,
quietly on his own. His work remained almost locked away until every-

thing had been put into its proper place. Without rushing, without pomposity and without arrogance, he waited until the fruits of his mind matured."

In 1858, when he was already sixty-eight years old, Möbius started work on a memoir that he wanted to present to the French Académie des Sciences. It was to be his submission for a prize that the academy had announced as a reward for the best essay on the geometric theory of polyhedra. In the essay he discussed the one-sided strip. Things could have turned out differently had Möbius been fluent in French, but he struggled with the language and did not win the prize. As a consequence, the memoir was never published and discovered only after his death. Thus it was sheer serendipity that Möbius's name was to be lent everlasting fame. Listing was not so lucky. His paper, albeit published in Göttingen's learned journal, remained unnoticed.

While Listing was doing research on the calculus of position in Göttingen, simultaneously trying to keep his family afloat financially, and Möbius was observing the sky in Leipzig, a young mathematician in Tuscany by the name of Enrico Betti was fighting for the independence of his fatherland. It was 1848 and the citizens of the various Italian kingdoms rose against the Austrian empire. Betti, then twenty-five years old, did not share the distaste that Möbius had felt for the military at the same age. Betti voluntarily joined the army under Ottaviano Fabrizio Mossotti, his former professor of mathematics at the University of Pisa and now captain of the Tuscany University Battalion. Betti saw action as a corporal in the battle at Curtatone-Montanara, an uprising that proved unsuccessful. After demobilization Betti taught mathematics at secondary schools for a few years, traveled to Göttingen, Berlin, and Paris, lectured on higher algebra, and was eventually appointed to the chair of analysis and geometry at the University of Pisa.

In 1861, a year after his appointment, Tuscany and various kingdoms united to become the Kingdom of Italy. From then on, and until his death in 1892, the civic-minded Betti alternated academic life with political and administrative duties. He served as a member of parliament, rector of the University of Pisa, director of the Scuola Normale Superiore, undersecretary of state for education, all the while continuing his research. At various times he held the chairs of mathematical physics, celestial mechanics, analysis and geometry. Betti did substantial work in theoretical physics as well as in mathematics. A pioneer, he heralded the development from classical to modern algebra.

He also advanced the relatively new field of topology. His most important contribution was the development of a means of counting the connectivity of surfaces and bodies. These numbers, later named "Betti numbers" by Poincaré, specify the connectivity of objects of any dimension. They were developed in a paper entitled *Sopra gli spazi di un numero qualunque di dimensioni* (On spaces of any dimension), which Betti published in 1871 in the Italian *Annals of Pure and Applied Mathematics*.

The Betti numbers of an object describe its features, such as the number of its components or the number of its holes and cavities. An object has one more Betti number than it has dimensions. One-dimensional objects such as the circle have two Betti numbers, two-dimensional objects, such as the surfaces of balls, pretzels, and bagels, have three Betti numbers. (Recall that *solid* spheres and bagels are three-dimensional objects. The surface of spheres and bagels, i.e., their skins, are two-dimensional.) Intuitively, the k-th Betti number counts the object's k-dimensional connectivity or holes.

Mathematicians start not with the first, but with the zeroth Betti number, which might seem strange, but that is the way they usually count. The zeroth Betti number specifies how many components the object is composed of. If we consider the usual objects such as cubes, spheres, and cylinders, the zeroth Betti number is one. If the object consists of two detached pieces, the zeroth Betti number is two. Next comes the first Betti number. It specifies how many holes there are in the body. The first Betti numbers of a circle or a paper cut out like a ring of Saturn are one, because they have a hole. For a sphere the first Betti number is zero (no hole), for a solid bagel it is one (one hole), and for the surface of a bagel it is two because in addition to the obvious hole, there is also the tunnel inside the bagel. Two- and three-dimensional objects also have second Betti numbers. They indicate how many cavities are hidden inside the objects. A sphere's second Betti number is one, since there is a cavity inside; a ball's second Betti number is zero because it contains no cavity.

The interesting thing about Betti numbers is that they are invariant under deformation—as was the case with the Euler characteristic. It does not matter whether the sphere is egg-shaped, whether the cylinder is formed like a Coke bottle, or whether the bagel has bulges. Betti numbers remain constant, no matter if and how the objects are twisted and warped—but not torn. This was not known for sure, however, for nearly half a century. Only in 1915 did a certain James Waddell Alexander at Princeton University prove that Betti numbers are, in fact, topological invariants.

Alexander was the offspring of a well-known patrician family. He was born in 1888 in Sea Bright, New Jersey, the son of noted painter John White Alexander. His forefathers were well-known Princetonians, and a street as well as two buildings at Princeton University are named after them. James Alexander was brought up in New York and Paris, where his parents mingled with the crème de la crème of the artists' world of the turn of the nineteenth century. Claude Debussy, Oscar Wilde, Henry James, and Auguste Rodin were only some of the notables they counted among their acquaintances.

Given his family's connection to Princeton University, it was only fitting that Alexander completed both his undergraduate studies in mathematics and physics and his Ph.D. at his forefather's alma mater. He made a name for himself as a first-class student and as a fiery speaker for left-wing causes, which was to make him suspect during the McCarthy era. During the First World War he joined the army and was assigned as a lieutenant, and later as a captain, to the technical staff of the Ordnance

Department at the Aberdeen Proving Ground. Subsequently, he was posted to Europe, where he met Natalia Levitzkaya, a White Russian émigré who would become his wife.

After the war Alexander returned to Princeton to become assistant professor, then associate professor, and finally full professor of mathematics. Having inherited considerable wealth, he was able to negotiate a smaller teaching load at half the salary. Finally, in 1933, he moved to the famed Institute for Advanced Study, where he became Albert Einstein's and John von Neumann's colleague.

During most of his career, his research focused on topology. First he provided, together with his mentor at Princeton, Oscar Veblen (1880–1960), the logical underpinnings for Poincaré's sometimes not fully developed theory of analysis situs. Then, in 1915, he found the proof that Betti numbers are, in fact, topological invariants. Four years later, he discovered counterexamples to one of Poincaré's conjectures…not the one, however, that we will deal with in the coming chapters.

In the 1920s Alexander turned to a problem that another Alexander had puzzled over more than two millennia earlier. In 333 BCE, Alexander the Great was deliberating how to open the Gordian knot. Impatient as he was, he did not deliberate for too long and just chopped the knot in two. He was not the only one who did not give a hoot about the mathematical details of his (mis)deed. Scouts, mountaineers, fishermen, and sailors tie knots daily without caring about the higher mathematics involved. But some mathematicians felt uneasy with the brute-force solution to the Gordian knot problem, and in time, knot theory became a subdiscipline of topology. In particular, an erroneous theory of the famous Scottish physicist Lord Kelvin (1824–1907) prompted scientists to turn their attention to knots.

Toward the end of his career Kelvin believed that atoms consisted of fine tubes that became tied up with one another and then buzzed around the ether. The theory was generally accepted for about two decades before being proved erroneous. In the meantime, however, this mistaken belief had led Peter Tait (1831–1901), another Scottish physicist, to categorize all possible knots. (Mathematical knots are different from their more mundane cousins in that both free ends are connected to each other. In other words, knots in knot theory are always closed loops.)

A superficial classification would use the number of crossings of two strands as determinants. But this type of categorization does not account for the possibility that two different-looking knots could actually be the same, i.e., that one of them could be turned into the other one by picking, plucking, tugging, and pulling, but without cutting or untying their strands. We now see why knot theory is a part of topology: different-looking knots are not necessarily different. If one knot can be "deformed" into another without cutting and reattaching the strands, the two are considered identical.

Tait discovered this quite intuitively. In his scheme of classification he attempted to account only for truly different knots, so-called prime knots, which cannot be disassembled into further components. His classification was not without its errors, however, as Kenneth Perko, a New York lawyer, discovered in 1974. Working on his living room floor, he managed to turn one knot with ten crossings into another knot that had been listed as different by Tait.

Nowadays we know that there exists only one knot with three crossings, another with four crossings, and two with five crossings. Altogether there are 249 different knots with up to ten crossings. Beyond that the number of possibilities rises quickly. There are no less than 1,701,935 different knots with up to sixteen crossings.

The central question in mathematical knot theory was and remains whether two knots are different, or if one of them can be transformed into the other one without cutting and reattaching the strands. A related question asks whether a tuft that looks like a knot is in fact an "unknot," because it can be disentangled by manipulating the strings, without cutting them. The well-worn trick of miraculously unknotting a complicated-looking tangle of strings, which magicians use to their advantage as the astonished audience oohs and aahs, works only because the tangle was an unknot to start with.

Henceforth, mathematicians sought characteristic traits, topological invariants as we would now say, that could clearly and unambiguously be attributed to the various knots, thus making them distinguishable from one another. This is where James Alexander comes in. In 1923 he discovered polynomials, which turned out to be suitable for the classification of knots. If the polynomials are different, the corresponding knots are also different. His paper "Topological Invariants of Knots and Links" was published in 1928 in the *Transactions of the American Mathematical Society*. Unfortunately, it soon became apparent that the reverse does not hold true: Different knots may possess identical polynomials. Since then, variants of Alexander's polynomial were proposed: the Jones polynomial, the Conway polynomial, the inelegantly named HOMFLY polynomial. Other mathematicians developed different systems of classification, and still others seek a workable recipe for how to convert identical knots from one form into another, equivalent form.

Knot theory is an example of a mathematical theory that was developed before applications were even considered. Useful implementations of knot theory surfaced only with time. Molecular biologists, for example, study the ways in which the long and stringy forms of the DNA molecule wind and twist so as to fit into the nucleus of a cell. And in the 1970s and 1980s quantum physicists suggested string theory to make quantum mechanics and the force of gravity compatible. This theory says that elementary particles are tiny strings, crushed together in higher-dimensional spaces. Obviously, the strings get entangled, and thus knot theory found another application.

During World War II, Alexander worked as a civilian for the Office of Scientific Research and Development of the U.S. Army Air Force. After that, he gradually became a recluse. In 1948 he gave up his professorship at IAS and became a permanent member of the institute without pay. Even though he still had an office, he shunned most contacts with his colleagues. In 1951, partly because of the persecution of people with left-wing views by Senator Joseph McCarthy, Alexander retired and practically disappeared from view. One of his last public actions was the signing of a petition in 1954 in support of J. Robert Oppenheimer, the famous physicist and father of the Manhattan Project, who was also suspected of Communist sympathies.

In his younger years Alexander had been an avid alpinist, scaling more than two hundred peaks in Colorado and then in Europe. One ascent of Longs Peak in Colorado is named Alexander's chimney. He and his wife even bought a chalet near the French mountain resort of Chamonix in order to climb the French Alps every summer and to enjoy the haughty French society, into which they blended beautifully. Not content to limit his alpine feats to his vacations, Alexander took to climbing the Princeton buildings...from the outside. Often he reached his office, located on the top floor of the mathematics building, through the window. Sometimes he left by the same way, especially when an unwelcome visitor was waiting in front of his office.

Alexander was also a proficient skier and loved baseball. Unfortunately, he was forced to give up sports in later years due to the aftereffects of a polio attack. He started to pursue more sedentary pastimes such as music, photography, and amateur radio. The design of a radio receiver circuit is named after him. He owned a beautiful house in which he and his wife entertained friends, not only from among the Princeton faculty but also from the art and business worlds. The parties the couple threw, with Alexander dancing the tango, and waiters with trays threading their way through the guests, were legendary. After his wife's death in 1967, Alexander's health rapidly declined, and he succumbed to pneumonia in 1971.

He did not enjoy teaching and did not need to do so to support himself. Nevertheless, whenever he could be persuaded to teach a course, his lectures were acclaimed for their precision and style. He was kind to his students and quite noncompetitive toward his colleagues.

This was in contrast to his grouchy colleague at Princeton, the topologist Solomon Lefschetz (1884–1972), who never forgave Alexander for having been appointed to IAS instead of him. Lefschetz, an engineer originally from Russia, who turned to mathematics after he lost both of his hands in a laboratory accident, wrote the monograph "Topology" in 1930, thus introducing Listing's word to the English-speaking world. Veblen, Alexander, and Lefschetz formed a topological trio that made Princeton into the world's center for this young discipline.

Among the topologists across the Atlantic, the German-born Heinz Hopf (1894–1971), who taught at the Eidgenössische Technische Hochschule in Zurich, provides a glimpse into the world history playing out at the time. When he was informed in the late 1930s in a letter from an ostensible friend, the head of Hopf's student fraternity, that "with regret" he had to be kicked out of the association of former members because his father was Jewish and his wife's brother an anti-Nazi, Hopf drew the consequences: He applied for, and received, Swiss citizenship.

Chapter 6

From Copenhagen and Hamburg to Black Mountain, North Carolina

Topological ideas and notions were evident throughout Poincaré's early work—for example, in his study on the sets of solutions to differential equations. These sets form surfaces and hypersurfaces in higher-dimensional spaces, and Poincaré studied their forms and shapes with the emerging tools of algebraic topology. The aim was to reduce topological questions to abstract algebra by associating various algebraic invariants with topological spaces.

The first paper in which he specifically dealt with topology was just a short note, written in a by-the-way fashion. His preoccupation with the subject began in earnest with a long paper published in 1895 in the renowned publication of his alma mater, the *Journal de l'École Polytechnique*. It was entitled *"Analysis situs"* and—like much of his often hastily written work—was not quite mature, not quite polished, and quite incomplete. Poincaré realized this himself and filled in some of the gaps with a *"Complément à l'analysis situs"* four years later. But he did not stop there. Over the next years, between 1900 and 1904, he published on average nearly one follow-up a year, for a total of five *compléments*.

Today, the development of algebraic topology is considered one of Poincaré's most brilliant achievements, but at the beginning of the twentieth century this work was mostly ignored. Eulogies after his death stressed his contributions to traditional areas of mathematics and physics, with the three-body problem being the most prominent example. One is hard-pressed to find more than passing mention of topology. In a forty-four-page appraisal of Poincaré's life and work, published a few months after his death, Viscount Robert d'Adhémar, a mathematician of some distinction at the beginning of the twentieth century, devoted no more than fourteen lines to the subject. Since the study of this subject matter required "finely honed intuition about space" and demanded "an enormous cerebral effort," he doubted that Poincaré's series of papers on analysis situs had found more than just a few readers. Émile Picard, on the other hand, a colleague much esteemed by Poincaré, pointed out the significance of analysis situs for the theory of functions and hailed Poincaré's work on it as one of his most important achievements. Picard had done much of the trailblazing work himself and was one of Poincaré's few contemporaries who could appreciate the new theory's significance. But even he did not devote more than one line to this newly emerging branch of mathematics.

At the centenary of Poincaré's birth, celebrated in 1954 with pomp and splendor throughout Paris, France, and Europe, the celebrated mathematician Jacques Hadamard did not even mention topology in his keynote address, *"Henri Poincaré et les Mathématiques,"* at the Sorbonne. And the equally renowned Gaston Julia devoted just five lines to this branch of mathematics in his commemorative speech *"Henri Poincaré, sa vie et son oeuvre"* (Henri Poincaré, his life and work). Finally, the transcript of the talk *"Poincaré et la Topologie,"* which the Russian mathematician Paul Alexandrov gave in The Hague, in the Netherlands, was omitted from the centenary volume altogether, even though every other detail of the festivities was fastidiously recorded.

The attitude changed only gradually. "In these days the angel of topology and the devil of abstract algebra fight for the soul of every individual discipline of mathematics," the German mathematician Herman Weyl remarked. Algebraic topology began to be fully appreciated by the mathematics community only in the 1940s. In his pathbreaking monograph, *Treatise on Topology*, Solomon Lefschetz wrote, "Perhaps on no branch of mathematics did Poincaré lay his stamp more indelibly than on topology." Since then the number of papers and books has skyrocketed. Google.com shows nearly 800,000 entries on "algebraic topology," and Amazon.com lists more than 120 books with "algebraic topology" in their title. Mathematics could not even be imagined today without it. But the importance of topology is not limited to mathematics. One hundred years after Poincaré's pathbreaking papers, applications can be found in such diverse areas as computer graphics, economics, dynamical systems, condensed-matter physics, biology, robotics, chemistry, cosmology, materials science, population modeling, and other fields of science and engineering.

The era of algebraic topology can be said to have begun on Monday, October 31, 1892. Félix Joseph Henri de Lacaze-Duthiers—an eminent zoologist and the discoverer of the Mediterranean mollusks that provided the purple-blue dyes mentioned in the Bible—presided over this day's session of the Académie des Sciences in Paris. Various communications on chemistry, physics, astronomy, mechanics, botany, geology, and physiology were read before the members, as was a report by the foreign minister about the observation of a lunar rainbow by France's ambassador to Armenia. At half past four, a secret committee convened to elect Paul Émile Appell, Poincaré's former fellow student in Nancy and the runner-up for the Oscar Prize, as a new member of the Académie's section of geometry. At a quarter past five o'clock the session ended.

One of the items on this day's agenda had been a lecture by Henri Poincaré entitled *"Sur l'Analysis situs."* It was a short talk, and the paper, subsequently published in the *Comptes Rendus de l'Académie des Sciences* (Proceedings of the Academy of Sciences), comprised no more than three printed pages. Poincaré read his paper at a time when he was still mourning his father, who had succumbed to the aftereffects of an unfortunate fall only five weeks previously. Coincidentally, it was just ten weeks after Enrico Betti had died.

The lecture started unspectacularly. "One knows of the important role that a surface's order of connection plays in the theory of functions, even though the notion is borrowed from a wholly different branch of mathematics, namely the geometry of situation or analysis situs," Poincaré told the members of the Académie. But he felt somewhat uneasy, in spite of his conviction of the subject's importance. Apparently he thought that he needed to legitimize his preoccupation with such an esoteric field. "People who shy away from geometry of more than three dimensions may think of such results as mere games of the mind," he admitted, but added that "because such research can have applications outside of geometry, it is of some interest to pursue it and extend the results to spaces of more than three dimensions." To further justify his choice of subject, he pointed out that Émile Picard's judicious use of Betti numbers had already shown the applicability of this new theory's tools to pure analysis and to ordinary geometry.

Then Poincaré quickly got to the point. "One may ask if—from the point of view of analysis situs—Betti numbers suffice to characterize a closed surface." It is true, he explained, that in three dimensions one can pass from one closed surface to another by continuous deformations, as long as the two surfaces have identical Betti numbers. Hence, these numbers are all that is needed to fully describe, topologically, a two-dimensional surface floating in three-dimensional space. From this, one could be led to believe that this is also true in higher dimensions. But is it really? At this point in his lecture, Poincaré pulled the rabbit out of the hat: "*C'est le contraire qui est vrai,*" he exclaimed—the opposite is true—and then went on to prove this by constructing a counterexample in four-dimensional space. And as we all know, a counterexample suffices to disprove a mathematical statement.

With this short piece—a note, rather than a paper—the stage was set for the things to come. But even though Poincaré considered analysis situs an extremely important subject, he himself did not seem to put much stock in it. When asked in 1901 by the Swedish mathematician Gösta Mittag-Leffler to summarize his work up to that point in his life, he devoted no more than 4 pages, out of 103, to analysis situs. (Moreover, these four pages are absent in the eleven-volume collected works of Poincaré.) Of course, by then only the first two *compléments* had been written, but even so one might have expected him to give more weight to this then-still-new field. The apparent lack of enthusiasm did not mean, however, that Poincaré disregarded the impact of topology. To the contrary, he saw it everywhere. "All the diverse paths with which I was engaged eventually led to analysis situs," he wrote in the assessment of his work. The paths he referred to included curves defined by differential equations, the three-body problem, two-variable functions, multiple integrals, perturbation functions, the theory of groups. "It is for these reasons that I devoted a fairly long work to this branch of mathematics."

And long it was. The *Journal de l'École Polytechnique*, the school's scholarly periodical, had been founded in year III after the French Revolution (i.e., in 1795) and had just completed its first hundred years of publication. It was time to start afresh, and Poincaré's paper covered no less than the first 121 pages of the new series' introductory volume. He regretted the paper's length but claimed that he could not help himself. "When I tried to hold back, I became unintelligible. So I preferred to be a bit talkative."

Poincaré started his paper by stressing the importance of geometrical figures for the study of mathematics. They are of great utility in the theory of functions of two variables, he asserts and reminds the reader how much one regrets their absence when studying functions of four variables. Wherein lies their contribution? Poincaré asserts, that scrutinizing figures compensates for the limitations of man's mind by calling on his senses. Furthermore, he states, "Geometry is the art of reasoning well with badly made figures."

"Badly made" must not be taken too literally. If they are not to lead the reader astray, figures must not shift the relative positions of their elements. But their proportions may be grossly altered. We recognize the standard definition of topology: stretching and squashing is permissible, but tearing and gluing is not. We also see that Poincaré's approach to algebraic topology was intuitive.

Unfortunately, intuition often goes hand in hand with lack of rigor. Nowadays, nobody would accept a less than completely strict proof because mathematics is nothing if not rigorous. One could be tempted to demand less rigor when dealing with geometrical objects. After all, everyone can imagine them floating around in space. But the small community of topologists should have known better, even as early as the 1890s, because they had been forewarned. In work dealing with two-dimensional objects, a subject where mathematicians may have believed that they could not be misled or deceived by intuition, mistakes were found that well-known mathematicians had made in papers written years earlier.

But the mathematics community was willing to forgive Poincaré's exuberance. While geometry, number theory, algebra, and calculus were already well-established fields, and no self-respecting mathematician would have accepted proofs by hand-waving, intuition was tacitly pardoned, if not completely accepted, in this new discipline. It would take another two decades for uncompromising rigor to be applied also to topology.

Rigorous proofs do not appear in Poincaré's seminal paper, nor do precise definitions. Often one has to second-guess what the author had in mind by interpreting the context of his writings. It turned out that *"Analysis situs,"* detailed as it was, represented more of an outline for further research than anything else. But whoever deigned to peer beyond its shortcomings soon realized that it was chock-full of new findings. "In almost every section is an original idea," the renowned French mathematician Jean Dieudonné exclaimed, enamored, nearly a century later. Poincaré's extraordinary intuition and imaginative powers rarely led him astray.

The paper starts out by defining spaces, subspaces, and manifolds. The latter can be imagined, for example, as flying carpets floating in the sky. (Throughout this book, I will use the words *object, space, body,* and *manifold* interchangeably.) Flying carpets correspond, to use technical language, to bounded, two-dimensional manifolds embedded in three-dimensional space. Floating quilts would also qualify because two-dimensional manifolds can be obtained by gluing together flat objects. Möbius strips are also manifolds—as they are created by gluing flat strips to themselves with a twist—and so are cylinders and even spheres. It is more difficult to imagine a sphere, but recall that topology doesn't see corners, so a sphere is created by attaching six squares together as a cube, or four triangles as a tetrahedron.

Poincaré also explains the notion of homeomorphism: Two shapes are said to be homeomorphic, or topologically equivalent, if one can be deformed into the other by pulling and creasing and crumpling, without tearing and gluing. A carpet is homeomorphic to a quilt but not to a poncho.

Poincaré presents many more concepts, notions, and terms in his paper, such as homology, one-sided manifolds (Möbius strips and their higher-dimensional counterparts), fundamental groups, simplices, complexes, simply connected complexes—a virtual menagerie of exotic abstract objects. I will introduce and explain them whenever they are needed throughout this book. Now I will dwell on just two of these concepts: Betti numbers and something called the duality theorem.

Poincaré claims that mathematicians have known and used the duality theorem in the past, albeit without proof. "This theorem has not been, I believe, ever stated. It is, however, known to many, who have even found some applications for it." Unfortunately, he did not give away who the "many" were and proceeds to give what he thought was a proof of the theorem. In the previous chapter I wrote that Betti numbers describe the shape of manifolds. The k-th Betti number counts an n-dimensional object's k-dimensional connectivity.

The duality theorem states that for closed manifolds the k-th Betti number and the $(n-k)$-th Betti number are identical. This can be verified, for example, for a circle (Betti numbers: 1,1), for a sphere (1,0,1), for a bagel (1,2,1), or for a three-dimensional bagel floating in four-dimensional space (1,3,3,1). The astonishing implication of this theorem is that the orders of connection of such manifolds in various dimensions are not independent of each other but must conform to certain rules.

"*Analysis situs*" became the Bible of algebraic topology. But in contrast to Holy Scripture one expects proofs in mathematical essays, not just prophecies and edifying stories. In the late 1980s, Dieudonné gave an in-depth summary and analysis of every section of "*Analysis situs.*" He did not mince words where he felt criticism was deserved. "Never bothered to show," "unformulated assumption," "totally unconvincing," "quite obscure," "sweeping assertion," "without any justification," and "incomplete" are some of his comments. Surprisingly, Dieudonné was nevertheless full of admiration and praise. He finished his round-by-round commentary with the words "Thus ends this fascinating and exasperating paper, which, in spite of its shortcomings, contains the germs of most of the developments of [algebraic topology] during the next 30 years."

At the time, not everybody saw "*Analysis situs*" in that light, however. Shortly after it was published, a Danish doctoral student in mathematics took a closer look. Poul Heegaard (1871–1948) had received his M.S. at the University of Copenhagen in 1893. As a schoolboy, Heegaard had not seemed predestined for mathematics. Until he entered high school, he believed that seven plus eight equals seventeen and he had to use his fingers to add numbers. But little by little, it turned out that he had quite a talent for abstract mathematics. After graduating from high school, he studied at the University of Copenhagen, even though he admitted that he never felt any burning interest for mathematics.

After finishing his undergraduate degree, he received a scholarship for travel abroad. His first stop was Paris. He found the French rather standoffish and did not profit much in terms of mathematics. Equipped with a letter of introduction from his professor in Denmark, he called on Professor Darboux, the dean of the Faculté des Sciences, for a social call. Only Darboux, who would write Poincaré's obituary twenty years later, was not very social. Heegaard waited in the anteroom for forty-five minutes before a secretary told him he might as well leave, because *monsieur le professeur* had tossed the letter of introduction into the wastebasket.

His experiences as an auditor at various courses were not much better. He attended lectures by Émile Picard, which he described as nothing less than a theater performance. Greeted with polite applause by the students, Picard would enter the lecture hall preceded at a light trot by a gofer whose job it was to place a glass of water and pieces of sugar on the table in front of the blackboard. The lectures themselves were much less awe-inspiring than the grand entrance since they just covered, word for word, Picard's book *Leçons d'Analyse*. Camille Jordan's lectures at the Collège de France were no better. "He went through the proof sheets of his *Cours d'Analyse*, occasionally pausing to pencil in a correction." The one French mathematician whom Heegaard should really have met, or whose classes he should at least have attended, he bypassed completely. Late in life he would write, "I have always regretted that I accepted the advice not to attend lectures by Poincaré, who was claimed to be unintelligible. His very intuitive exposition has later been of great importance to me when I met it in printed books."

It was probably during his half-year stay in Paris that his interest in topology originated, even though there is no indication that he spent much time working on mathematics. Instead, he seems to have devoted most of his stay in Paris to satisfying an old desire of his: to learn Chinese, not exactly what comes to mind for most people visiting this beautiful city. In any case, he made up for lost opportunities during the second semester of his study year, which he spent in the German city of Göttingen.

The reception he received there was much kinder than the one in Paris. The Danish student was taken under the wings of Felix Klein, the doyen of the math department. Klein had Heegaard give a seminar talk and two lectures at meetings of the Mathematische Gesellschaft (the Mathematical Society). Heegaard found the atmosphere at the University of Göttingen stimulating and conducive to research. Unfortunately, his respect for German mathematicians would later translate itself into misguided support for the totalitarian Nazi regime.

Back in Copenhagen, Heegaard started teaching at various high schools. By holding down several positions simultaneously, he improved his financial situation sufficiently to allow him to finally marry Magdalena, his sweetheart from his student days. In the meantime he had started thinking about a doctoral dissertation and finally settled on topology. Unfortunately, he had not read Poincaré's recent paper on duality and was considering a problem in a similar vein. But some of his "well-meaning" colleagues mentioned it to him, barely hiding their glee. "It must be unpleasant for you that Poincaré has solved your problem," one of the older professors told him maliciously. Actually, this was the first Heegaard ever heard of *"Analysis situs."* He would have liked to read the paper, but the relevant issue of the *Journal de l'École Polytechnique* was not available at the public library. It circulated among the professors and Heegaard knew he would not get to see it for many weeks. Sitting on needles, he ordered the issue at his own expense from a bookseller. When it finally arrived, he leafed through it impatiently, and then it suddenly hit him. He noticed that Poincaré had made a grave error. The purported proof of the duality theorem was flawed.

For Heegaard this turned out to be a blessing: He had found his thesis topic. It was late fall, and the Christmas break during which he would have time to work was approaching. He set out to prepare a critique of Poincaré's proof of the duality theorem. This brash young man—just a schoolteacher with no more than a master's degree—deigned to attack the acknowledged master of European mathematics (with the possible exception of David Hilbert in Göttingen). It was the only time in Heegaard's life that his working style could be described as feverish. It took him no more than a few weeks to find a counterexample to the duality theorem.

Let us recall that according to the theorem, the k-th Betti numbers must be equal to the $(n-k)$-th Betti number. Heegaard constructed an example of a three-dimensional manifold—an intersection of a certain cone with a cylinder—whose Betti numbers are $(1,1,2,1)$. This contradicts the duality theorem. Finding a counterexample to a theorem can mean either that the counterexample is wrong, or that the theorem's proof is wrong, or that everything is based on a misunderstanding. In this case, it was the third alternative, as we shall see below.

In January 1898, Heegaard handed in a dissertation entitled *"Forstudier til en topologisk Teori for de algebraiske Fladers Sammenhaeng"* (Preliminary Studies of a Theory of Connectivity for Algebraic Surfaces). The thesis was accepted and Heegaard received his doctoral degree. He promptly sent a copy of the thesis to Poincaré. Poincaré was taken aback. Too honest a scientist, he could not dismiss this challenge as simply the work of a newcomer. Realizing that something was seriously amiss with his proof but unable to read the Danish thesis, he asked Heegaard for clarifications. A correspondence followed in which Heegaard translated parts of his paper into French. Once everything was clear, Poincaré acknowledged that that his own proof fell seriously short.

To resolve the apparent contradiction between his duality theorem and Heegaard's *très remarquable* (very noteworthy) Ph.D. thesis, Poincaré wrote the first *complément*. Maybe a small detail was missing and the proof could be repaired? Not so, according to Heegaard. "Not only has the theorem not been proven, it cannot be true," he declared. After all, he had found a counterexample.

Poincaré's *"Complément à l'analysis situs"* was published in 1899 in the journal of an Italian mathematical society, the Circolo Matematica di Palermo. This mathematical association had been founded fifteen years earlier, in 1884, by the wealthy mathematician Giovanni Guccia, who provided the meeting place, a library, and all necessary funds to publish the *Rendiconti del Circolo Matematica di Palermo* (Conference Proceedings of the Mathematical Circle of Palermo). The learned journal had a high reputation, and some of the most important mathematical papers of the early twentieth century appeared in it. Unfortunately, the Circolo, which had close to a thousand members in its heyday and boasted eminent mathematicians from France and Germany on the editorial board of the *Rendiconti*, lost its lustre after Guccia's death at the outbreak of the First World War.

After taking a close look at his original paper, Poincaré concluded, to what must have been his immense relief—ten years after the Oscar fiasco—that this time there was nothing really wrong with the proof. It was just that he had proved something different from what he had thought. This came to him when he went through Heegaard's counterexample with a fine-tooth comb. Painstakingly redoing all calculations, he realized that there are two different sorts of Betti numbers. In *"Analysis situs"* he had thought, like everybody else, of an object's Betti numbers as its connectivity in higher dimensions, just as described in chapter 5. But when he actually proved the duality theorem, it was for another set of numbers, which he henceforth called reduced Betti numbers. Heegaard, correctly assuming that Poincaré meant the connectivity numbers, had concluded that the proof was wrong. In what was probably the academic understatement of the year, Poincaré admitted that his proof "has a feeble point."

The confusion that had arisen in the wake of Heegaard's criticism ruined Poincaré's contention about the connectivity of objects in space. Poincaré's primary objective in writing the first *complément* had been to provide another proof of the duality theorem for reduced Betti numbers. But the five follow-up papers gave him occasion to clean up his act, thereby setting the stage for the future development of algebraic topology.

After publishing his first *complément* to clarify his duality theorem, he reworked much of his earlier ideas and simplified them in his second *complément*, published in 1900 in the *Proceedings of the London Mathematical Society*. Otherwise it contained not much new information, except for a significant statement, about which I will have more to say in the next chapter. The third *complément* was published in the *Bulletin de la Société de Mathématique de France* in 1902, and the fourth appeared in the same year in the *Journal de Mathématiques pures et appliqués*

(Journal of Pure and Applied Mathematics). The fifth *complément* was published in 1904, again in the *Rendiconti*. In this final *complément*, in the very last lines, Poincaré stated the famous question that is the subject matter of the present book.

Meanwhile Heegaard's dissertation became well-known all over Europe in spite of having been written in Danish. All over Europe, that is, except in Denmark, where it was ridiculed and deemed worthless by Heegaard's malevolent colleagues, some of whom had apparently not even read it. But mathematicians who had studied his work thought highly of the thesis. The French Mathematical Society even had a translation published in its learned journal, the *Bulletin de la Société Mathématique de France*, albeit eighteen years after the Danish version had appeared. This did not happen quite without misgivings, however. In the preface, the editors of the *Bulletin* asked themselves whether it was really appropriate to publicize the work of this Scandinavian, at a time when nations were particularly jealous about the reputation of their scientists. Their answer was that by going ahead with the translation they were acting in the true spirit of Poincaré, who never cared about such things. They justified themselves further by pointing out that Heegaard's thesis was being published in French (under the title *"Sur l'Analysis situs"*) in order to heighten the public's awareness of Poincaré's achievements and to further augment his glory. At the same time, they cautioned the reader not to always take Heegaard's trenchant criticism at face value.

With a dissertation that had evoked international attention, Heegaard had got off to a promising start. But he had a wife and six children to care for and a university salary would not suffice to keep them in comfortable circumstances. To make ends meet, he passed up a university career for the time being and instead took on various tasks and teaching jobs at naval academies and military schools. The grueling teaching load—eight hours a day, six days a week—left no time for anything else. Given this, and his lack of a burning desire to occupy himself with mathematics, Heegaard did very little research. On the upside, his lack of involvement with university matters left him unaffected by strife and disputes with fellow mathematicians.

An occasion to return to academic work arose when Heegaard was asked to write an article on topology for the *Enzyklopädie der mathematischen Wissenschaften* (Encyclopedia of Mathematical Sciences). This reference work was originally planned as a simple dictionary. But in preparatory talks, Felix Klein proposed that the encyclopedia provide an overall picture of the position of mathematics in general culture. Eventually, the project developed into a gigantic work that would span four decades, between 1898 and 1935, and comprise twenty volumes.

Heegaard accepted the invitation and got as far as writing an outline and a bibliography. But when he started on the introduction, he was already overwhelmed by his workload. In addition, he had got into time-consuming squabbles with his colleagues again. Therefore he asked the encyclopedia's editor, Friedrich Wilhelm Franz Meyer, professor at the University of Königsberg and himself an early topologist, for an assistant. Soon the German mathematician Max Dehn, who had just recently received his Ph.D., was recruited to assist Heegaard with his task.

Heegaard's junior by seven years, Dehn hailed from an assimilated Jewish doctor's family in Hamburg. The parents and eight children did not think of themselves as Jewish and had even converted to Protestantism. Dehn received a broad education at the *Gymnasium* in Hamburg, before studying mathematics in Freiburg and Göttingen. He obtained his doctorate in 1899 under David Hilbert. Shortly after Hilbert posed his famous twenty-three problems at the International Mathematical Congress in Paris in 1900, Dehn solved number 3 on the list. This feat— the first of the twenty-three problems to be solved—earned him the position of a lecturer at the University of Münster. When the First World War broke out, the good German patriot Max Dehn—by then already a thirty-seven-year-old professor at the University of Breslau—joined the army. He participated in half a dozen battles and was decorated for his military service with the Ehrenkreuz für Frontkämpfer (Honorary Cross for Front Fighters).

Dehn and Heegaard had met at a conference in the German city of Kassel in 1903. On the train back, the two men talked about the foundational problems of topology. Dehn suggested an axiomatic approach to this new mathematical science: Postulate as few axioms as possible and let the rest flow from there. He found a kindred spirit in Heegaard, who enthusiastically concurred. After the two men had decided to cooperate on the article for the encyclopedia, Heegaard visited Dehn in the summer of 1905 to "initiate him to his viewpoints."

The *Enzyklopädie* entry appeared in 1907. Even though Heegaard had initially regarded Dehn as just an assistant, the footnote to the seventy-page article indicated otherwise. Heegaard had done the preliminary study of the literature, it said, and had contributed an essential part of the article, while Dehn had the responsibility for the final form.

The encyclopedia entry represented a new approach to topology. Starting with the most basic notions and facts, everything else was deduced purely by logic. In spite of their predilection for the axiomatic approach, Dehn and Heegaard were well aware of the importance of descriptiveness. After introducing the axioms, they included a section, entitled *Das Anschauungssubstrat* (the visual substrate), in which they encouraged the reader to use imagination to envisage those parts from the theory that could be visualized in three dimensions. It is through such visual interpretations that topology acquires its value, they stated.

Their appeal to spatial imagination notwithstanding, not everybody liked Dehn and Heegaard's axiomatic approach. Felix Klein thought it was too abstract. He regretted the lack of intuitive descriptiveness that he always favored and recommended. To understand anything from the encyclopedia entry, he wrote, the reader would already have to be intimately acquainted with the field.

The *Enzyklopädie* article became influential throughout Europe, and Heegaard's standing within the mathematics community began to rise again. In 1910 the position of professor of mathematics at the University of Copenhagen became vacant. Heegaard was urged by his friends to apply, but he felt unqualified for the job. Moreover, it paid much less than his current teaching jobs. But the friends insisted, and when their entreaties had no effect, they worked on his mother until he relented. Heegaard was a gifted teacher and the students liked him. But he did not get along well with his colleagues. As was Heegaard's wont, squabbles and disagreements with colleagues followed, and seven years later he resigned his post amidst a row about appointments that even hit the daily press. Infighting with peers and superiors finally also led to his resignation from the Danish Mathematical Society. Heegaard's version of the events that led to his resignation are not known because the crucial pages 103 and 104 of his handwritten notes were inexplicably removed from his autobiographical record.

A year later, he received a call to the University of Kristiania in Norway (later renamed the University of Oslo), where he stayed for twenty-three years, until 1941. He again did little scientific work, except for helping in the publication of the works of the famous Norwegian mathematician Sophus Lie (1842–1899). He assisted in the editing of two volumes before the effort became too much for him. Heegaard preferred to travel the country giving popular talks on science throughout Norway. He even went as far as a town called Kautokeino in the extreme north of Norway, necessitating a 400-mile overland voyage, followed by 700 miles on a boat, and then a trek inland for 150 miles, part of it on a reindeer sledge. His lecture in Kautokeino was translated into Lappish by the local shopkeeper.

There was one further, and ultimately futile, attempt at academic work. In the spring of 1946, after he had long retired from the University of Oslo, he wrote an excited letter to his friend Jakob Nielsen in Denmark. "Hold on to your hat! I believe I have solved the four color problem!!" The four-color problem was one of those notorious conjectures that everybody tried, and failed, to prove. It stated that any geographical map could be painted with no more than four colors such that no two adjacent countries are of the same color. Heegaard made plans to have the manuscript translated into English so that it could be published in an American journal. The rude awakening came soon thereafter. "This is only to tell you that you can throw the manuscript in the wastebasket," a deeply disappointed Heegaard wrote Nielsen. He had found a mistake in his purported proof and thus entered the long list of frustrated hopefuls who thought they had proved the Fermat Conjecture, the Riemann Conjecture, and, oh, yes, the Poincaré Conjecture.

In the meantime the Germans had invaded Norway. Regretfully, Heegaard was not ill-disposed toward the Nazi occupiers of his new homeland. Indeed, he seems to have found a comfortable arrangement with the new situation. The authorities even permitted him to broadcast a series of popular talks on science over Norwegian radio. To appreciate the significance of this, it must be pointed out that only members of the Norwegian Nazi Party were allowed to own radio sets. As a Nazi sympathizer, Heegaard found his reputation sullied, so much so that hardly any obituaries appeared after his death in 1948. In a centenary volume of the Norwegian Academy of Science in 1960, Poul Heegaard was mentioned as someone "whose sympathies definitely lay with the authoritarian states."

Max Dehn's further career was quite the opposite of Heegaard's. When the Nazis came to power, Dehn, who had always thought of himself as a German Protestant, began to realize what it meant to be Jewish. He had been named professor at the University of Frankfurt in 1921, as the successor to Ludwig Bieberbach, who was later to become a notorious anti-Semite and Nazi. With the enactment of the heinous Gesetz zur Wiederherstellung des Berufsbeamtentums (Law for the Restoration of the Professional Civil Service) in April 1933, Jews were excluded from academic positions. As a war veteran and carrier of the Ehrenkreuz, Dehn was initially exempted, but the respite was short. In 1935 he lost his professorship for good. He was wise enough to send his children abroad, a son to America and two daughters to England.

In the infamous Kristallnacht in November 1938, Dehn was arrested but then set free for the remainder of the night because the prisons were overcrowded. Realizing that he would be rearrested the next day, Dehn and his wife immediately left Frankfurt. They fled to nearby Bad Homburg, where they were put up by a courageous colleague. A few weeks later they fled Germany, first to Copenhagen and then to Trondheim in Norway. For a year and a half, Dehn managed to eke out a living, lecturing at the Technical University of Trondheim. But the relative security came to an abrupt end when Germany invaded Norway in April 1940. Dehn was kept in hiding at a Norwegian farm for a few months while Scandinavian colleagues arranged for his and his wife's flight. In early 1941, at a time when his coauthor Heegaard lived comfortably in Oslo, the Dehns made their tortuous and fearsome way over the Norwegian border and then on via Finland, Russia, Siberia, Japan, and the Pacific to the United States.

The deluge of exceptional European mathematicians led to a paucity of available academic positions in America. Dehn wandered around the United States until he finally found permanent employment at Black Mountain College in North Carolina in 1945. This institution, a nonaccredited college of sorts, was more of an educational commune than an academic establishment. Students and faculty were expected to roll up their sleeves to build campus facilities and grow crops for food. Dehn received $40 a month plus room and board. In spite of his mathematical isolation, he seems to have enjoyed his time at Black Mountain. The breadth of his knowledge and culture was such that he was able to teach not only mathematics, of which he was the only faculty member at the college, but also Latin, Greek, and philosophy. Max Dehn died in 1952.

His mathematical results and techniques are used frequently to this day, particularly Dehn surgery (used to adapt three-dimensional manifolds), Dehn's Lemma (which was actually only proven by Kneser in 1929), and Dehn's Algorithm (which is used in group theory). He is also remembered for his work in knot theory, most notably for his proof that left and right trefoil knots are not equivalent.

* * *

Poincaré died prematurely at fifty-eight. Ever since his submission for the Prize of King Oscar, he had been preoccupied with a proof of nothing less than the stability of our universe. His final paper on the subject was entitled simply *"Sur un théorème de géométrie"* (On a geometrical theorem). It was published in 1912, also in the *Rendiconti*. It was untypical work for Poincaré, and the year of publication says it all: It was written only a few months before Poincaré's untimely death.

As has been pointed out, Poincaré's papers were often not quite mature, not quite polished, and not quite error-free. But Poincaré never knowingly published unfinished work. This time he made an exception. *Je n'ai jamais présenté au public un travail aussi inachevé* (I have never presented to the public a work as unfinished as this one) are the opening words. He then recounts that he had spent many months on this problem without being able to solve it. It would have been best, he says, to have given the matter a rest for a few years and then started over. "This would be fine if I were sure that I could pick it up again one day; *mais à mon âge je ne puis en répondre*" (but at my age I am not sure I will be able to rise to the occasion). The subject's importance was so great in his opinion, however, that he felt that his work so far should not have been in vain. Was it a premonition of his coming demise that led him to submit his unfinished contribution to the editors of the *Rendiconti*? After all, he had suffered medical problems four years earlier, at the International Congress of Mathematicians in Rome, and was now due to have surgery. Poincaré sent his paper for publication. A few months later he was dead.

This important theorem had to do with the three-body problem, the partial solution of which had won him the Oscar Prize nearly a quarter of a century before. At that time, he had shown the existence of periodic orbits for three bodies circling one another. The proof left much to be desired, however, because it held only for bodies of small masses. But the universe contains large masses, and it would be nice to know whether massive planets also stay on their orbits. Pondering the problem, Poincaré concluded that the answer depended on whether a certain geometric conjecture was true or false. For two years he tried out many special cases and found that in each of them the conjecture held. But a general proof eluded him. "My conviction that the theorem is true grew day by day but I was incapable of putting it on a firm foundation."

To state the conjecture is simple. Consider a ring-shaped region bounded by two concentric circles. Now transform that region so that the points on the inner circle advance in the clockwise direction, the points on the outer circle advance in the counterclockwise direction, and everything in between is twisted in a smooth fashion. Poincaré claimed that there are at least two points in the region that stay at rest. The situation calls to mind the eye of a hurricane, where everything is quiet while all surroundings are in commotion. This is what the stability of our universe should depend on.

Happily for our universe, Poincaré was proven right. It took George David Birkhoff, then an assistant professor at Harvard, only a few months to prove that, in fact, there are always two points that do not move. Birkhoff's proof was published in the *Transactions of the American Mathematical Society* in January 1913 and established him as the foremost American mathematician of the early twentieth century.

Chapter 7

What the Conjecture Is Really All About

Toward the end of the nineteenth century the main thrust of the then-still-young discipline of topology was the classification of bodies and spaces. The aim was to group together all bodies that are topologically identical, i.e., that have shapes that can be morphed into one another by stretching and squeezing, but not by tearing and gluing. Clearly, trial by error—deforming a shape until it fits the other one—would not be an efficient way of verifying whether two shapes are equivalent. It would be much worse than searching through all the possible configurations of Rubik's Cube. After all, the combinations of Rubik's Cube are large but finite, while the morphing possibilities are endless.

So during the eighteenth and nineteenth centuries, the idea arose to attach labels to the bodies. Once this is done, it would only be a matter of comparing the two objects' labels. If they are identical, the two objects can be morphed into each other. If the labels are different, one would be certain that the two objects are topologically different.

In technical terms, labels were sought that do not vary when the body is transformed into a different but topologically equivalent shape. Bodies that belong to the same group should be assigned the same label. What remained was to find an appropriate labeling scheme. What is it that characterizes a body? This question has kept mathematicians busy until this day. Clearly, a single number, such as the number of the body's corners, would not do since a ball and a bagel have zero corners and would thus be classed in the same group even though they are topologically different, while a cube would be classed as different from a pyramid, even though one can be morphed into the other.

We saw how the Swiss mathematician Leonhard Euler had made a first step toward the objective in the eighteenth century. He found an equation whose solution remains invariant for any convex, three-dimensional body. Later, his countryman Simon l'Huilier showed that various kinds of holes, cavities, and annular faces can compensate for one another, thus giving the same result even if the bodies are differently shaped and not necessarily convex. Hence the Euler-l'Huilier equation is essentially of no value for classification purposes since it classes everything into a single basket.

Seeking a new kind of label, Poincaré decided to try a different tack: Describe the body's topological characteristic not by one number, but by a series of numbers. The numbers he thought of were the Betti numbers and the so-called torsion coefficients. In chapter 5, I explained that the Betti numbers are the number of holes a body has; the torsion coefficients are a bit more tricky to describe, which is why I won't do it here. Regardless, both the Betti numbers and the torsion coefficients are part and parcel of the body's so-called homology groups. This is not an easy concept at all, so let me just state here that each body has a collection of Betti numbers and torsion coefficients, and that bodies with different collections of these numbers are topologically different.

After having published "Analysis situs" in 1895, and a first set of corrigenda and addenda four years later, Poincaré followed up with another set of clarifications in 1900, the "Second complément à l'Analysis situs." As pointed out in chapter 6, it did not treat much new material and was written mainly to simplify and clarify previous results.

However, it did contain a bombshell, which Poincaré saved for the very end. The paper concluded with the significant claim that "every polyhedron all of whose Betti numbers are equal to one, and which is free of torsion, is simply connected." In modern language, this statement would read, "A three-dimensional manifold that has the same homology groups as the three-dimensional sphere is homeomorphic to it."

Both versions of Poincaré's statement are technically precise but not very helpful to nonspecialists. Before we rephrase them in a more understandable way, the notion of torsion needs to be explained. In a somewhat abstract sense, torsion coefficients describe a body's twistedness. Take the one-sided Möbius strip. (If you don't have one handy, you can create one by cutting out a strip of paper and attaching the ends to each other, after turning one of the ends through a half circle.) Obviously, it is a round object, somewhat resembling a circle. Now, starting at any point on the edge, run your finger around the strip until you return to the starting point. You will touch all points on the object's boundary, so your finger will have performed a cycle. But to trace the boundary, you went around twice. This indicates that the body is, in a sense, twisted.

So let me rephrase Poincaré's statement in simple, if not quite precise, words: Any body that contains no holes and is not twisted can be morphed into a sphere. (Both the body and the sphere are three-dimensional objects, floating in four-dimensional space.) Poincaré did not provide a proof. The paper had already run to thirty-three printed pages, and he did not want to stretch it out any more. So he closed with the words "In order not to make this work even longer, I will here just announce the theorem whose proof still needs some elaboration."

To say that the proof required some elaboration may have been the scientific understatement of the outgoing nineteenth century. In fact, the bold statement was no more than a cheeky guess. But it carried an eerie resemblance to a similar claim made three centuries earlier. "I have discovered a truly remarkable proof which this margin is too small to contain," the French judge Pierre de Fermat had written in his copy of Diophantus's *Arithmetika* in 1630, and thus putting the world in suspense for more than three and a half centuries. Like Fermat's claim, Poincaré's announcement was premature. Unlike Fermat's claim, it was wrong.

The somewhat cocky Poincaré apparently thought he had a proof in his pocket, but as on numerous other occasions, he was mistaken. Fortunately, he realized this himself. After several years of valiant but unsuccessful efforts at proving the statement, he began to suspect that something was seriously amiss with his purported theorem.

Poincaré began to ask himself whether bodies that possess the same Betti numbers and torsion coefficients as spheres must, indeed, be spheres. It gradually dawned on him that this need not be so. *Nous allons voir qu'il n'en est pas toujours ainsi, et pour cela nous nous bornerons à donner un exemple,* he wrote in the fifth *complément,* in 1904 (We shall see that this is not always the case and in order to show this we limit ourselves to giving an example).

The example, more precisely the counterexample to his statement, is quite an elaborate construct that became famous and is nowadays called Poincaré's homology sphere, aka Poincaré's dodecahedral space, aka Poincaré's icosahedral manifold. It is a three-dimensional body. But don't get your hopes up, one cannot make a mental picture of it. That's because the body is suspended in four-dimensional space.

This is a good occasion to remind ourselves once again of what it actually means for a body to be two- or three-dimensional. A disk is a two-dimensional object. So is a balloon. (We just consider the skin of the balloon, not the inside.) A disk lying on a tabletop is a two-dimensional body embedded in a two-dimensional plane. This is a special case, however, and to fully visualize a two-dimensional body one often needs to go to three dimensions. The disk floating in the air, the balloon, the surface of a ball, the crust of a bagel are two-dimensional objects embedded in three-dimensional space.

A solid ball and a solid bagel, on the other hand, are three-dimensional bodies. Again, to visualize the full spectrum of three-dimensional bodies one often has to go to higher dimensions. Even the Klein bottle, which is, after all, nothing but a two-dimensional surface, must be embedded in four-dimensional space so that its features can be fully exhibited. Don't worry if you cannot make a mental picture of that. Some people, such as the physicist Roger Penrose, claim they can think in four dimensions, but most people cannot. However, one can gain an understanding by appealing to analogy: The three-dimensional sphere is to the ball what the surface of the ball is to a disk.

So what does Poincaré's homology sphere look like? I have already mentioned that it is a three-dimensional object floating in four-dimensional space. There are various ways to describe, if not to depict, it. One way of getting a partial picture of a higher-dimensional object is to project it down into lower dimensions. This is like shining a lamp at it and observing its shadow on the opposite wall. A balloon, for example, throws a round shadow onto a surface. Creatures living in a two-dimensional world would have to surmise from the shadow on the floor what the two-dimensional object looks like in three-dimensional space above. The disk, in turn, may throw a straight-line shadow, and the straight line may throw a pointlike shadow. Of course, a round shadow could also be thrown onto a surface by a cylinder, so the shadow does not reveal everything.

Another way to evoke a higher-dimensional object is to describe how to construct it. For example, one could describe the torus (the hollow bagel) to two-dimensional creatures via the following instructions: "Take a sheet of paper, glue the two opposing edges together to form a cylinder, then glue the two openings of the cylinder together to create the torus." Even though a two-dimensional creature could not visualize the resulting object, she would be able to grasp it.

This is the approach we will take to get some notion of what Poincaré's homology sphere looks like. There are quite a few different ways to construct the sphere, and a well-known paper lists eight different methods. And there are more. The one discovered by the master himself involved two double tori (that's two hollow bagels with two holes each) that are glued together in a certain way. Five years later, Max Dehn suggested a "surgery" technique that sounds suspiciously like medical malpractice but has actually become quite standard in mathematics. He suggested extracting a pretzel-like object from the three-dimensional sphere and sewing it back differently. Then he showed that the resulting Frankenstein monster was actually Poincaré's homology sphere. Of course, the surgery performed by Dehn is purely virtual, taking place in four-dimensional space.

We will describe a different construction method that was devised in 1929 by Hellmuth Kneser (1898–1973). Before we embark on his method, let me describe the man. The son of the mathematician Adolf Kneser, Hellmuth Kneser received his Ph.D. in 1921 under David Hilbert in Göttingen. He obtained a chair in Greifswald in 1925 and was called to the University of Tübingen twelve years later. During the Second World War, Kneser let himself be swept along by the Nazi atmosphere, even though he is believed to have been more of a conservative than a full-blooded Nazi. Nevertheless, he became a member of the motorized Sturmabteilung, the notorious SA, and toyed with the idea of joining the official Nazi party NSDAP because he approved of their nationalist actions and feelings. He signed his letters "Heil Hitler" and supported the racist ideas of Ludwig Bieberbach, the notorious Nazi mathematician and founder of the purely Aryan journal *Deutsche Mathematik* (German Mathematics). Because of his rather minor role during the war, Kneser never suffered any adverse consequences for his objectionable attitude. After the war, he assisted in the founding of the Mathematical Research Institute in Oberwolfach, which would become one of the world's foremost research institutes in mathematics. Kneser was named its scientific director for a short period in 1958.

To construct the homology sphere, Kneser prescribed the following procedure. Take a dodecahedron, the solid that is bounded by twelve regular pentagons. Two pentagons always lie on opposite sides of the dodecahedron, so there are six pairs. Identify one pair, stretch and bend the dodecahedron so that—after twisting it by one-fifth of a revolution (seventy-two degrees)—the two pentagons can be glued together. Okay, so there is some twisting involved, but it's more of a rotation in order to align the edges and corners, rather than the cut-twist-and-glue type of twisting of the Möbius strip that Poincaré explicitly excluded. Now do the same thing with the other five pairs of pentagons, and voilà, you have the homology sphere. After all this stretching and bending and aligning, you may understand why a higher dimension is needed to accommodate the resulting object; three dimensions do not suffice to even picture it.

If that does not make complete sense, maybe the following will. It is well-known—among algebraic topologists, that is—that the Poincaré homology sphere is the boundary of a compact, simply connected four-manifold in five-dimensional space, with the second Betti number equal to one. That really does clarify things, does it not?

After the homology sphere is constructed, one still has to establish in what way it is a counterexample. This is exactly what Poincaré did in the paper. Length was no more a consideration, and at sixty-six pages the fifth *complément* turned out twice as long as the second. Poincaré computed the Betti numbers and the torsion coefficients and found them to be equal to the numbers and coefficients of the ordinary three-dimensional ball floating in four-dimensional space. But the deformations— elongating, bending, turning, and gluing—turned the dodecahedron into a shape that cannot be morphed into a sphere. Thus, even though Poincaré's body has the Betti numbers and torsion coefficients of a sphere, it can't be morphed into one.

To this day, Poincaré's homology sphere is the only known three-dimensional, untwisted, holeless body that is not topologically equivalent to a three-dimensional sphere. It may well be that it is the only counterexample that exists. Hence one could mistakenly surmise that Poincaré was nearly correct in his bold statement. After all, what is one counterexample among so many instances where his statement holds? But in mathematics there is no such thing as nearly correct or approximately right. Absolute truth is what one is after, and one counterexample suffices to prove a whole theorem wrong.

Having found a counterexample to his initial claim, Poincaré realized that even a series of numbers, such as the torsion coefficients and the Betti numbers, would be unable to furnish a classification of spaces and bodies. Hence, a different kind of topological invariant was needed, and Poincaré started looking elsewhere for characteristic attributes that would be appropriate to classify bodies into groups. The concept he eventually thought of to establish a labeling system for bodies and shapes came from algebra: the so-called fundamental group.

Before explaining what the fundamental group is, let me point out Poincaré's dilemma. Designating a body's fundamental group as its label was an appealing possibility. It was an ingenious insight to introduce algebraic expressions, instead of numbers, into what was until then thought of as a wholly geometric subject. But he had already tried something similar with the homological group (which produce the Betti numbers and torsion coefficients) and it had not worked. Would the fundamental group do any better?

To keep things simple we start with two dimensions. Let's consider a sphere—say, a balloon or a basketball. Take any point on the surface of the ball—for example, the outlet of the valve. (Remember that the surface of a ball is two-dimensional. The ball itself is three-dimensional.) We will call that point the tee. Attach an orange rubber band to the tee and loop it around the ball. Now attach a black rubber band to the tee and also loop it around the ball any way you want. You can slide the bands about—you can push, pull, stretch, and shrink them—until they are aligned next to each other. (Actually they should cover each other.) You can do this with any two rubber bands, irrespectively of how they were strung around the ball originally. In addition, you can do one more thing the significance of which will become apparent later. By tightening the bands more and more, you can shrink them until they are no more than a single point at the tee.

Now consider a torus, say a hollow bagel, and decide on a tee somewhere on the bagel's surface. Take a blue rubber band and, starting at the tee, loop it around the bagel's outside. Now take a yellow rubber band and, again starting at the tee, loop it around the bagel, but this time go through the hole. (You'll have to cut the band, string it through the hole, then join the two ends together.) These two rubber bands can also be slid about the surface of the bagel, but they will never become aligned next to each other. Hence for bagels there are two kinds of loops: blue loops that go around the bagel, and yellow loops that go through the hole. Any two blue loops can be aligned and any two yellow loops can be aligned. But a blue loop can never be aligned with a yellow band. Also, none of the blue or yellow loops can be shrunk to a point. The yellow loop gets caught at the edge of the hole, which prevents further shrinking; the blue loop would have to cross the hole, which is not legitimate. (The rubber bands are only allowed to slide along the surface, not hang in midair.) However, one can envisage a third kind of band; let us say, colored orange. It is attached to the tee and slid onto the bagel sideways. The orange band can be shrunk to a single point at the tee.

There are also loops that wrap both around the bagel and through the hole, and loops that wrap twice around the bagel and once through the hole, and loops that wrap twice through the hole and once around, and so on. Could it be that the strange collection of loops—the ones that shrink, the ones that capture the crucial holes, and the infinitely many other loops that seem to be combinations of them—can be used to describe the space? This is in fact the essence of Poincaré's Conjecture in three dimensions. But in higher dimensions one needs more.

Let us leave rubber bands behind for a little while and turn to something new. I will describe what mathematicians understand by a *group*. In general parlance a group is a collection of objects that have something in common as, for example, a group of people (they are all humans), a group of groupies (they all admire the same pop star), a group of objects on the table (they lie close to one another). For mathematicians, however, this is only part of the story. For them a group is a notion that is of paramount importance in algebra. It consists, first, of a collection of objects, and second, of a mathematical operation that combines two of the objects. For example, the objects could be the integers 1, 2, 3..., and the mathematical operation could be the addition of two numbers.

But not every collection of objects together with an operation represents a mathematical group. For a collection of objects together with an operation to qualify as a group, four conditions, let's call them the groupie requirements, must be satisfied. First, the result of the operation that combines two members of the group must also be a member of the group. Integers and the operation of addition fulfill this requirement since, for example, 3 plus 4 equals 7, which is also an integer. Even numbers also meet the requirement, because the sum of two even numbers is even. On the other hand, odd numbers do not, since the sum of two odd numbers is not odd.

The next groupie requirement is that when the operation is performed twice in a row, the order in which the operations are performed does not matter. Addition satisfies this requirement because $(3+5)+7$ is the same as $3+(5+7)$. (It does not matter if one adds 3 and 5, then adds 7 to the result, or if one adds 5 and 7, and only then adds 3.) Mathematicians call this requirement the associative property.

The third requirement is that the group contain a neutral element (the so-called identity element), which—when combined with any element of the group—does not change the result. Clearly for the operation of addition, the neutral element is the number zero, since adding zero to any number leaves the number unchanged ($4+0 = 4$).

The final groupie requirement is that for each element the group also contains an inverse, such that combining an element and its inverse produces the neutral element. For integers, the inverse is the negative of the number—e.g., $5+(-5)$ equals zero, the neutral element.

We have already seen that integers form a group under the operation of addition. On the other hand, odd numbers do not form a group under subtraction ($9 - 5$ equals an even number), nor do integers under division ($9/2$ is not an integer). Neither do the real numbers under the operation of multiplication: The neutral element in this case is 1, and the inverse of x is $1/x$, but there is no inverse to the number zero.

The smallest group one can imagine under the operation of addition is the group consisting only of the number zero. Since $0+0 = 0$, it contains one element its inverse and the neutral element, all in one guise. This group is fairly uninteresting, though very important to mathematicians. They called it the trivial group.

So far our examples involved integers, real numbers, even and odd numbers, and zero. But groups do not necessarily have to consist of numbers. For example, we could consider roads in the United States as members of a group. The operation is "driving a round-trip starting and ending in New York City." So the elements would be the round-trips from NYC to Philadelphia, from NYC to Miami, from NYC to San Francisco, etc. To ascertain that such trips do indeed form a group, we must verify that the four groupie requirements hold. It may help to have a map of the United States in front of you while reading the following paragraphs.

The first requirement holds since any two elements can be combined. The round-trip NYC–Philadelphia can be combined with the round-trip NYC–San Francisco. Since the whole journey starts and ends in the Big Apple, it also represents a round-trip and hence is a member of the group. Let us prove that the second groupie requirement is satisfied, which states that it does not matter in which order the combining is done. Combining the round-trip NYC–Philadelphia with the NYC–San Francisco round-trip, then adding the round-trip NYC–Miami represents the same journey as combining NYC–Philadelphia to NYC–Miami and then adding NYC–San Francisco.

What is the identity element? It is quite simply your coveted parking spot in Manhattan. Proof: Combining the round-trip NYC–Philadelphia with the parking spot is the same as the round-trip without the parking spot. And what about the inverse element? It is the same trip driven in the opposite direction. Proof: Driving from NYC via Philadelphia to Boston and back to NYC, then immediately continuing the journey by driving from NYC via Boston to Philadelphia and back to NYC is—algebraically speaking—the same as never leaving your parking spot. Wasted time, effort, and gasoline play no role in math. Hence the round-trips from NYC, together with a parking spot in Manhattan, form a group under the operation of driving. As they say in Latin, *quod erat demonstrandum.*

Now we come to the crux of the matter. The rubber bands around the balls, bagels, and pretzels can also be considered mathematical groups. In stark contrast to the trivial group, mathematicians find the group made up of the rubber bands so important that they called it the fundamental group. How do these bands form a group? Let's see.

We will go straight to the bagel and deal with the basketball later. The bagel lies flat on the table and two rubber bands are attached to the tee again. But this time they need not be wound around the bagel only once. The blue band can be wound around the bagel several times, and the yellow band can be threaded through the hole more than once. Furthermore, the blue band can be wound around the bagel from left to right or from right to left (i.e., clockwise and counterclockwise). And the yellow band can go through the hole from the top down or from the bottom up.

Hence the winding operations can be described by two numbers. The first number indicates how often the blue band is wound around the outside of the bagel; the second tells us how often the yellow band goes through the hole. The sign before the numbers indicates the direction in which the winding is done: Plus could indicate right to left and top down, minus could indicate left to right and bottom up. Hence a pair of integers, with the appropriate signs, fully describes the windings that have been performed. A pair of numbers such as (5, −3) means that the band is wound five times around the outside from right to left, and three times through the hole from the bottom up.

And now, lo and behold, we have a group. Its elements are the number of times the bands are looped and threaded, and the operation is the addition. To see that this is a legitimate group, we must verify the four groupie requirements. Let us add the winding (−2, 2) to the winding (5, −3). Looping the blue band twice from left to right and five times from right to left is the same as three loops from right to left, i.e., −2+5 = 3.

Threading the yellow band through the hole twice from top to bottom and three times from bottom to top is the same as threading it once from bottom to top, i.e., $2 - 3 = -1$. Hence we obtain the pair of integers $(3, -1)$, which is also a winding.

The second requirement is that the order in which windings are added is of no importance. It is simple, if a little tedious, to show that this requirement is fulfilled, so I'll leave it out. What is the neutral element? It is the "no-winding," which is denoted as $(0,0)$. It corresponds to the orange band that was slid on sideways and shrunk to a point at the tee. Adding a no-winding to any winding does not change the original winding.

What about the inverse elements? For this we seek an unwinding that, when added to a winding, produces a no-winding. That's an easy one. Adding the winding $(-5, 3)$ to the winding $(5, -3)$ equals $(0, 0)$, which is the no-winding.

Thus we have proved that the rubber bands on a bagel, expressed as a set of two integers, form a group under the operation of addition. It is the fundamental group of the torus. Similarly, the fundamental group of a figure 8 can be described by the combination of three integers, which indicate how often and in which direction each of the blue, yellow, and green bands is wound around the figure 8 and through its holes. The pretzel's fundamental group consists of sets of four numbers, such as $(3, -2, -12, 6)$. Its neutral element is $(0, 0, 0, 0)$, which corresponds to the orange band that is neither wound around the pretzel nor threaded through any of its holes but just slid onto the pretzel sideways.

Returning to the basketball, what is its fundamental group? The ball allows only one kind of rubber band, the orange one that can be shrunk to a single point. Hence its fundamental group, consisting only of the neutral element, is the trivial group. As it turns out, even the three-dimensional sphere studied by Poincaré has a trivial fundamental group.

When dealing with rubber bands we limited ourselves to loops that are one-dimensional. But there are also loops of higher dimensions, as we will now see. I will illustrate this in two dimensions.

Let us recall that the familiar one-dimensional loops start and end at the base point, which we called the tee. The loop can also be envisaged as a line segment bent and stretched, such that both ends reach the tee. Moving to two dimensions, we need to consider a square instead of the line segment. Now bend and twist the square's surface into some kind of shape, the requirement being that all points on the square's four sides end up at the tee. Mathematically speaking, this is called a mapping. The result will look somewhat like a warped parachute: Interior points of the square are mapped to the canopy and the lines that descend from it, while the points on the square's four sides are mapped to the parachutist's backpack.

The square can be mapped to different parachutes. One can envisage this as the three or four parachutes that are attached to heavy military equipment when it is dropped from a transport aircraft or to the space shuttle when it is brought back to earth. All parachutes are attached at the same spot to the hardware. Some of them can be made to overlap each other exactly by pushing and pulling and stretching and squeezing the canopies; they are called equivalent. But not all parachutes can be so arranged. As was the case with rubber bands, where loops around the bagel could not be aligned with loops through the bagel's hole, some parachutes cannot be made to overlap either. They get stuck in higher-dimensional space at the brims of higher-dimensional holes. We may then have several classes of parachutes: All parachutes that are equivalent are said to belong to the same class.

It was shown above that the rubber bands, i.e., the one-dimensional loops, form something that is called the fundamental group. The classes of two-dimensional loops, i.e., the parachutes, also form a group. It is called the second homotopy group (the fundamental group being the first homotopy group). And homotopy groups do not end there. As mathematicians are wont to do, they immediately generalized to third, fourth, and higher homotopy groups.

The story so far can be summarized by saying that the fundamental groups convey information about the topological shape—the hole structure—of two-dimensional bodies floating in three-dimensional space. In an analogous fashion, higher homotopy groups describe the higher-dimensional holes of higher-dimensional bodies. Bodies with different homotopy groups cannot be morphed into each other.

While expounding on rubber bands and parachutes, I have occasionally glossed over an important point. A bagel is a three-dimensional object, as we can easily verify by biting into it. But when talking about bagels in the previous paragraphs, I was actually referring not to the bagel itself but to its surface. And a surface, as we know, is two-dimensional. So the rubber bands that we have been discussing are one-dimensional loops winding around two-dimensional objects that are floating in three-dimensional space.

Then we moved to parachutes, thereby increasing all relevant dimensions by one: The canopies represent two-dimensional loops that wind around three-dimensional objects that float in four-dimensional space. Everything clear? If it is not, then keep in mind that part of what is beautiful about topology is the unique way it boggles the mind.

With the introduction of the fundamental group, Poincaré definitely incorporated algebra into the study of topological objects. We already saw that coffee cups and bagels have the same fundamental groups; so do bodies shaped like the figure 8 and two-handled cups, pretzels and three-handled cups, etc. The fundamental group is a topological invariant: It remains the same even while the body is squeezed, pulled, shoved, and pushed. Is the opposite also true? If two bodies have the same fundamental group, must they be topologically equivalent?

Poincaré was not sure, but we now know that the answer is an emphatic no. Two bodies may have the same fundamental group and still be topologically different. Only seven years after Poincaré's death, in 1919, James Alexander came up with counterexamples. He was able to prove that structures exist that cannot be morphed one into the other, even though they have the same fundamental group. They are three-dimensional objects called lens spaces that are constructed by gluing two bagels together in a certain fashion in four-dimensional space. For such bodies the fundamental groups do not suffice as a labeling system. More delicate invariants must be sought to distinguish the bodies.

But fundamental groups were too good a tool to be dismissed so easily. After all, lens spaces are very special bodies. Maybe they could be disregarded and fundamental groups could be used as labels for all other bodies? Or maybe a class of objects exists about whom the fundamental group does provide all necessary information? Speculating on such questions was what led Poincaré to formulate the question that would later come to be known as his famous conjecture.

Spheres have a trivial fundamental group. If the fundamental group is not good enough as a general classification system, maybe at least all bodies that have the trivial fundamental group are topologically equivalent to the sphere? Poincaré actually thought so. But after his cocky announcement four years earlier, he had become more cautious. This time he did not phrase his hunch as a theorem. He introduced the last paragraph of the fifth *complément* rather coyly with the sentence *"Il resterait une question à traiter"* (There remains a question to be dealt with), then goes on, "Is it possible that the fundamental group of a manifold be trivial and yet the manifold not be homeomorphic to a sphere?"

This is it, the famous conjecture. Actually, it was no conjecture at all as yet, but just a seemingly innocent question. But the way in which it was phrased makes it obvious that Poincaré thought that the correct answer would be no. To make it more descriptive, let us rephrase the question as the conjecture as it has became known, and let us do that in terms of bodies with rubber bands strung and looped around it. "Three-dimensional bodies, all of whose rubber bands can be shrunk to a point, can be morphed into a sphere." So Poincaré suspected that all that was needed to recognize a three-dimensional sphere was information about one-dimensional loops.

But this time Poincaré was not going to stick out his neck. He ends his paper with a well-worn phrase that scientists have used innumerable times over the ages as an excuse for not dealing with a problem to which they do not know the answer. "*Mais cette question nous entraînerait trop loin*" (But this question would lead us too far astray).

Leaving the question unanswered, Poincaré mailed his paper to the editorial office of the Circolo Matematica at 30 Via Ruggiero Settimo in Palermo on November 3, 1903. It was published a few months later, in 1904.

Framing his hunch as a question rather than as a theorem was a stroke of genius. Now the onus of answering and providing the necessary proof was not on him but on his colleagues. Little did Poincaré know that his unanswered question would keep generations of mathematicians busy. In fact, as the following chapters will show, throughout the twentieth century scores of mathematicians spent large parts of their careers on the conjecture. Poincaré apparently thought that the answer to the question would be positive, but this was by no means certain. For two-dimensional objects it obviously is: Rubber bands around balloons, eggs, cubes, and pyramids can be shrunk to single points, and the objects can be deformed into spheres. (Recall that by *two-dimensional objects* we mean the shell of the egg and the outsides of the cubes and pyramids.) But of course one cannot deduce from this that the conjecture holds for any dimension.

Publications on the subject became so voluminous that the American Mathematical Society had to set aside an entire subject classification number for Poincaré's Conjecture. Initially, some attempts were made at finding counterexamples. They all proved unsuccessful, and eventually everybody became convinced, like Poincaré himself, that the conjecture was correct. All that was now needed was a proof.

And for a hundred years, mathematicians from all over the world searched....

Chapter 8

Dead Ends and a Mysterious Disease

The first person to take a serious crack at the Poincaré Conjecture was the Englishman John H. C. Whitehead, who usually went by his middle name Henry. Henry Whitehead's father was an Anglican minister; his mother was Isobel Duncan, one of the few female math scholars at Oxford at the time. Mathematics seemed to run in the family: Alfred North Whitehead, the famous philosopher and coauthor— with Bertrand Russell—of *Principia Mathematica*, was his father's brother.

Whitehead's life was quite unspectacular, with little for the biographer to chronicle, as a colleague would put it in an obituary. Henry was born in Madras (now Chennai) in India, where his father served as a bishop. Apparently the parents felt that India was not the ideal place to bring up an English child, and when he was one and a half years old, they brought him back to Britain. There he was raised by his maternal grandmother, who lived in Oxford. Later in life he would fondly remember the drives in her carriage, which shaped his lifelong attachment to this university town. Only after his father's retirement, fifteen years later, did the boy see his parents more than just occasionally.

Henry got the best education that the English school system had to offer. Even though he was of above average intelligence according to his teachers in primary school, he was no child prodigy. Somewhat careless in his work and not very good at mathematical manipulations, he nevertheless managed to pass the entrance examination to Eton, the most prestigious of England's boys' schools. After Eton he did his undergraduate studies at Balliol College at Oxford. Known for his high spirits and good humor, he was considered what the English would call a jolly good fellow. Excelling above all at boxing, cricket, and billiards, he also did fairly well academically, winning first-class honors in Moderations (a first set of examinations) and Finals. But by no means was he an outstanding student. So upon graduation, it seemed natural for him to look for a job, and he moved to the City to become a stockbroker.

Financial markets did not provide the environment that Whitehead wished for in a career. He did not suffer life among the London banks and brokerage houses for long; after barely a year in the City, Whitehead returned to Oxford to do more work in mathematics. Fortunately, one of the world's foremost mathematicians, Oswald Veblen, from Princeton University, was then on a sabbatical visit to Oxford. Together the two men would produce some major works in differential geometry. When Veblen's year at Oxford was up, Whitehead—who had just received a Commonwealth Fellowship—followed him back to Princeton. During the ensuing three years his interest in and talent for mathematics were firmly established. Together with Veblen he wrote *The Foundations of Differential Geometry*, which became a classic in its field, before his interests started shifting toward topology.

Back at Oxford, Whitehead met and fell in love with Barbara Sheila Carew Smyth, a concert pianist with an interest in agriculture and husbandry. The two married in 1934 and had two sons. World War II found Whitehead at legendary Bletchley Park. Supervised by Alan Turing, one of the fathers of modern computing theory, he spent four years cracking German ciphers. During the night of the worst blitz on London, he took shelter in the wine cellar of a friend and passed the time working on a mathematical problem. Somewhat surprisingly, not a single bottle was opened that night. In 1947 he was named Waynflete Professor of Pure Mathematics at Oxford's Magdalen College.

Upon the death of his mother in 1953, Whitehead inherited some cattle from her estate, and he and his wife established Manor Farm in the village of Noke, eight kilometers north of Oxford. The couple liked to entertain friends and students in an informal atmosphere. Whitehead was universally liked, with everybody calling him by his first (actually middle) name, Henry. His habit of breaking into song at appropriate— and sometimes inappropriate—occasions made him the life of any party but proved to be somewhat disconcerting to his hosts at more formal events. Whitehead died on a sabbatical visit to Princeton quite unexpectedly from a heart attack. It happened at eight o'clock one morning while he was walking back from an all-night undergraduate poker game.

On both sides of the Atlantic he gained the reputation of a profound and deep thinker in mathematics. In the late 1950s he founded the journal *Topology*. (Recently the board of editors collectively resigned because the subscription price had become too high.) But he did not dismiss the lighter sides of intellectual exercise, and at one time or another he was fond of card puzzles and of palindromes—"step on no pets" being one of the latter of his own making. By all accounts he was an inspiring teacher and a wonderful talker. As a writer he was less polished, since he had little time and not much taste for elegance. "He gave his readers a rough ride," one of his friends admitted. As a lecturer, he was even worse. There were many jokes about his style of lecturing. One of them claimed that the initials of his name, J.H.C., stood for "Jesus, he's confusing." It was his cheerful personality that most impressed colleagues and students alike. "The fact that the Mathematical Institute at Oxford...is one of the happiest of all the university departments and laboratories anywhere, is largely a tribute to his gaiety and wholesomeness of spirit," the *Times* of London wrote in an obituary.

Whitehead was eight when Poincaré died, and about thirty when he turned his attention to the as yet not very famous conjecture. At that time it was just one of a host of open problems; nobody knew how fiendishly difficult the proof would be. Hence, Whitehead simply turned to the standard tools of algebraic topology in search of a proof.

Imagine a balloon that is sitting on the floor and prick it at its zenith without letting out the air. If you project each point of the balloon onto a plane, the punctured balloon can be mapped to the two-dimensional plane. The removed point, the zenith, corresponds to infinity. It was already known that a punctured balloon is contractible, which implies that any loop on it can smoothly be shrunk to a single point. Recall that this is precisely what Poincaré's Conjecture is about: bodies whose loops can be shrunk to a single point. The idea of removing a point from a body and mapping the remainder can be extended to three dimensions. Thus, a punctured three-dimensional sphere can be opened up onto the three-dimensional world in which we live.

Whitehead starts with a closed three-dimensional space—the surface of a four-dimensional object floating in four-dimensional space—and hopes to prove that if it has a trivial fundamental group, it is a three-dimensional sphere. He punctures a hole in it and notes that this new open space is still three-dimensional and contractible. He then proves that a contractible open three-dimensional space must be the Euclidean three-dimensional space that we live in. Finally, he adds the point back at infinity, closing up the ball to get the desired three-dimensional sphere. QED.

Alas, the proof was not to be. To make a long story short, Whitehead's argument was wrong. But let us recount the series of events as they happened. Whitehead wrote down his proof and on August 1, 1934, submitted the paper to the *Quarterly Journal of Mathematics*, a publication of Oxford University Press. Maybe because of his prominence in Oxford's mathematical circles, the paper with the bland title "Certain theorems about three-dimensional manifolds (I)" was published post-haste in the same year. But Whitehead could not bask in the glory for long. Within a few months he realized he had made an error. The sinking feeling one gets in one's stomach with the realization of a published error is not to be wished on anybody. Forget the wise words to the effect that every false attempt somehow furthers human knowledge and that errors are necessary for progress. Wrong is wrong. And forget, of course, "Certain theorems about three-dimensional manifolds (II)," which Whitehead had promised in paper number I.

So on February 8, 1935, Whitehead submitted a recantation that was published in the following *Quarterly*. It was only nineteen lines long and carried the meek title "Three-dimensional manifolds (Corrigendum)." He introduced the rectification with the words "Theorem 1 in my recent paper, on which the other theorems depend, is false." One can just feel the pain with which this sentence must have been written. Even some tardiness on the part of the *Quarterly* could have saved Whitehead from the embarrassment. After all, a paper by another author that was submitted just three weeks after Whitehead's faulty proof had to wait until the following year to see the light of day. If only Whitehead's work had not been rushed into print.

But after the corrigendum's opening sentence Whitehead redeemed himself big-time. In the remaining eighteen lines he constructed a counterexample to his original claim that is so intricate and elaborate that he may be forgiven for not having thought of it immediately. It involves two bagels, one of which is wound around and linked to itself inside the other, like the link of a chain. The resulting configuration, called the Whitehead link, is only the beginning of the story, however. Next, he envisaged a series of Whitehead links, judiciously nested inside one another. Each is wrapped around and linked to itself within the previous one, ad infinitum. The open three-dimensional object created in this manner, which came to be called the Whitehead manifold, is actually contractible: The loops (and even parachutes) just pass through the whole object, unaffected by the endless sequence of links, until they contract to a point. This open space, however, is not homeomorphic to our usual three-dimensional Euclidean space and cannot be pulled, squashed, or twisted into a sphere by gluing in a point at infinity. So...no proof of Poincaré's Conjecture. End of story.

So maybe the Whitehead manifold is a counterexample to the Poincaré Conjecture? Unfortunately, it also falls short in this regard. A counterexample would have to be a closed three-dimensional space that is not a sphere, even though its loops contract to points. But since a point is missing, the Whitehead manifold is open, not closed. One might think of just adding a point at infinity to make Whitehead's manifold a closed space. However, this cannot be done because the manifold gets more and more twisted as one travels through the successive links toward infinity; it cannot be tied together with a single point. So...no counterexample to the Poincaré Conjecture either.

Being the first example of a three-dimensional contractible object that could not be deformed into a sphere, the Whitehead manifold had interesting implications for algebraic topology. Such objects had never before been imagined. But once the door was open, many more were found. In fact, it was later proved that there exist uncountably many of them. So Whitehead's false attempt did further human knowledge after all. But prove Poincaré's Conjecture it did not.

Another attempt on Poincaré's Conjecture was made a quarter of a century later by a relatively obscure man from Greece, Christos Papakyriakopoulos. The mathematician with the eight-syllable name—which is usually abbreviated to a simple Papa—was born in 1914, one of two children of a well-to-do family in Athens. His father was a textile merchant. At age eighteen, Papa enrolled in the prestigious National Metsovion Institute of Technology (now the National Technical University of Athens), where one of his professors convinced him to channel his efforts toward mathematics. This he did, choosing algebraic topology as his preferred subject. However, there was nobody in Greece who could teach him, and he had to study the textbooks on his own.

Starting in 1935, he got swept up in political events. King George II of Greece had been exiled twelve years earlier, and the Greek people were to decide in a referendum whether they wanted him back. Papa voted openly against the king's return, but the majority went the other way and the king returned to Athens.

When the Nazis occupied Greece in April 1941, Papa was busy working on his Ph.D. He completed the thesis in 1943. It was on the so-called *Hauptvermutung* (The Major Conjecture), formulated in 1908 and considered at the time to be a central problem of topology. The *Hauptvermutung* conjectured that any two triangulations of a polyhedron are equivalent, in the sense that any two triangulations of a manifold could be triangulated further and further until the resulting nets were equal. Papa proved that this seemingly simple, but difficult to prove, theorem was true for polyhedra of dimensions one and two. Later, the *Hauptvermutung* was also proven to be correct by Edwin Moise—more on him below—for dimension three. But the *Hauptvermutung* is false for general spaces in higher dimensions. So beware, dear reader: Not all famous conjectures are, in fact, correct. The thesis brought him to the attention of the most renowned of modern Greek mathematicians, Constantin Carathéodory, who was working in Munich at the time. Since nobody in Greece could judge the dissertation, the doctorate was conferred on Papa on the recommendation of Carathéodory.

King George II had fled to exile in Egypt when the Nazis invaded Greece, and with him large parts of the army had left the country. Papa's brother joined the Rimini Brigade, Greek troops that were loyal to the exiled government, and was killed in action in northern Italy while fighting with the Allies. In contrast, Papakyriakopoulos joined the Communist Party's National Liberation Front. When civil war broke out in Greece in 1944, Papa moved to the countryside with the Communist guerillas. He whiled away his time as an elementary-school teacher in the city of Karditsa, about three hundred kilometers from Athens.

During a lull in the fighting, Papa returned to the Institute of Technology in Athens. He had begun work on a theorem concerning loops in three-dimensional manifolds that had been left unfinished by Max Dehn and was on his way to becoming a serious mathematician. But the atmosphere at the institute was not friendly toward Communist sympathizers.

When the professor for whom he had worked as an unpaid assistant was fired, he had to start looking for other pastures.

Papa's work on three-dimensional manifolds served as an entry ticket to a new life. As soon as he felt that he had closed the gap in the proof that Dehn had left open, he sent the paper to the distinguished mathematician Ralph Fox at Princeton University. Fox quickly realized that this proof too was incomplete. But he was so impressed with this young man who produced work of the highest caliber in total isolation in Greece that he invited him to Princeton. Papa eagerly accepted. He arrived in Princeton in 1948, moved into a hotel, and never moved out. In fact he occupied the same room for the rest of his life, returning to Greece only once, in 1952, for the funeral of his father.

At first Papa was at the Institute for Advanced Study; later he moved to nearby Princeton University. When he was offered an academic position, he refused, not wanting to burden himself with teaching duties and administrative chores. It was not easy to approach him. He was quite secretive and would not share his ideas with anyone before he thought they were mature. His sole passion was research, and this he pursued to abandon. As an Alfred P. Sloan Foundation Research Fellow, he was paid a small stipend, and he had a little money from his family. Fortunately, his austere lifestyle did not require much anyway. Papa never married, had no family and no ties to anyone outside his work. Since he did not teach, he had no students either. His only luxuries in life were a visit to the movies once a week and a nice office in Fine Hall, the legendary building in which Princeton University's math department was housed. He kept to a strict schedule, having breakfast in the student cafeteria at eight, starting work in his office by eight thirty, then doing research until eleven thirty, having lunch, usually alone, until twelve thirty, and tea in the common room of the math department at three. This was followed by a reading of *The New York Times*, seminars at four, and work again in his office until the evening.

During his first ten years in America, Papa produced proofs to three important open problems: the Loop Theorem, Dehn's Lemma, and the

Sphere Theorem. These problems were concerned with how loops, disks, and spheres can be embedded into three-dimensional manifolds. The papers Papa wrote on the subject were considered so outstanding that he was invited in 1958 to give a lecture at the International Congress of Mathematicians in Amsterdam titled "Some Problems on 3-Dimensional Manifolds." To be asked by the organizers of this congress, which takes place only once in four years, to address colleagues from all over the world is considered one of the greatest distinctions in the mathematical profession. Six years later Papa was the first recipient of one of the two Veblen Prizes in geometry, which have since then been awarded by the American Mathematical Society only once every five years.

Even though Papa had no teaching duties, he took an active interest in young colleagues. Joan Birman, a freshly minted Ph.D. from the Courant Institute in New York, recounts how she met Papa, then already a well-known personality in the mathematics community. At a seminar at Princeton she approached him and told him that she wanted to work on three-dimensional manifolds. Admitting that she knew little about the subject, she asked him for suggestions about what to read. Papa said that he would think about it. A week later, at the next seminar, he had a twelve-page list of papers ready, written out by hand, just for her. Birman, who would later become a world-renowned knot theorist and topologist and is now a professor emerita of Columbia University, still remembers how astonished and touched she had been at the time. This great man did not think it beneath his honor to do a literature search for a new Ph.D. who did not even know enough to ask interesting questions.

When the military junta in Greece fell in July 1975, Papa prepared a visit to his fatherland. He even had his passport renewed. But it was not to be. In the summer of 1976, at age sixty-two, he died of stomach cancer. (Is topology incompatible with longevity? Poincaré died at fifty-eight, Whitehead at fifty-six, and Papa at sixty-two.) The National Technical University of Athens established a prize in Papa's memory, to be awarded every year to an outstanding freshman in mathematics.

It is a reflection of his solitude that throughout his career Papa never had even a single coauthor. Altogether, he published fifteen papers. As CVs of mathematicians go, this is not considered a voluminous output, but—as attested by the Veblen Prize—his work was of the highest standard. Papa's life and personality inspired the Greek writer Apostolos Doxiadis to shape the main character in his best seller *Uncle Petros & Goldbach's Conjecture* partly after him. The author had visited Papa at his office in Princeton and described him as a sweet gentleman, withdrawn, reclusive, and otherworldly, but kind and gentle, who gave his visitors an old-world aristocratic feeling.

Given his somewhat esoteric character, it is not surprising that Papa devoted the last fifteen years of his life to one single aim: proving the Poincaré Conjecture. He took to it with a vengeance, and the conjecture kept him in its grip from the early 1960s until his death.

He had it well planned out. As a first step he was going to reduce the Poincaré Conjecture to something more manageable. He announced his intention in spring 1962 in a submission to the *Bulletin of the American Mathematical Society*. Comprising no more than seven pages, it was a concise and rather short paper. It described how Poincaré's topological conjecture could be reduced to a pair of two conjectures, one in topology and one in group theory. Prove the two new conjectures and Poincaré's Conjecture would be proven. Papa gave few details and no proofs in the article, promising them for a subsequent paper in another journal.

But then something happened. At the Courant Institute of Mathematical Sciences, the home of New York University's math department, a young mathematician by the name of Bernard Maskit took a close look at Papa's proposal. Maskit was twenty-seven at the time and was working on his doctoral dissertation. His thesis adviser, Lipman Bers, a refugee from the USSR (now Latvia), asked him to study a paper from the previous century and to use modern topological techniques to prove its main results. The bright graduate student was successful, and Bers told his then son-in-law, Leon Ehrenpreis, at New York University, about this. Ehrenpreis, in turn, mentioned it to some people at Princeton, where it eventually came to the attention of Papa. Realizing that Maskit's work had implications for his own work, Papa sent his conjectures, again through Ehrenpreis, to this unknown young man at the Courant Institute. Maskit started reading Papa's paper. Soon the feeling befell him that something was wrong, and after a while his suspicion was confirmed. He managed to construct a counterexample to Papa's topological conjecture. Hence, any attempt to solve Poincaré's Conjecture by proving the two new conjectures was doomed from the start. Before Papa could really embark on his program, it had already been shown to be deficient.

At first, the situation seemed even worse than in Whitehead's case. At least there it had been the author himself who had found the error in his faulty proof. But Papa was not to be discouraged. After hearing of Maskit's counterexample, he went to work again to rectify his faulty arguments. He overcame the hurdle posed by Maskit with an adapted set of conjectures. But again Maskit constructed a counterexample. Papa adapted his conjectures once more, and Maskit tore them apart again. Altogether Maskit constructed three counterexamples, and Papa changed his conjectures each time to overcome the problems.

Finally the valiant attempts were successful: Papa came up with a new set of conjectures to which Poincaré's Conjecture could be reduced. This time neither Maskit nor anybody else was able to poke holes into his arguments. Maskit's challenge and Papa's new set of conjectures were published back-to-back in the *Bulletin of the AMS* in 1963. And as promised, Papa also published a fifty-six-page paper in the same year, giving all details and proofs, in the *Annals of Mathematics*.

The dates of submission of the papers present a somewhat unclear picture, however. Papa's initial announcement—with the incorrect formulation of the conjectures—was communicated to the *Bulletin of the AMS* on March 4, 1962. But two and a half months earlier, on December 22, 1961, he had already submitted the full paper—with the corrected conjectures—to the *Annals*. Why did he not rectify his announcement of March 1962 to the *Bulletin* if he was already aware of the error in December 1961? Or was December 22, 1961, the date of the original submission and a final version was received later? The delay between the submission to the *Annals* and the paper's eventual publication, fifteen months, could account for this. Today's journals routinely list all relevant dates—original submission, revisions, and final acceptance—to avoid any such confusion.

So Papa had achieved his first aim: Poincaré's Conjecture had been reduced to different, hopefully simpler conjectures in group theory. Now it was time to embark on the second stage, which was to prove the replacement conjectures. But here, Papa's efforts came to a grinding halt. He spent the next twelve years in futile attempts to make headway. Unfortunately, his penchant for secrecy did not help him. Everybody knew what he was working on, but since he never shared his ideas, nobody could offer any constructive criticism. So instead of realizing his mistakes and moving on, he wasted many years on an impossible dream.

His brave efforts were ultimately unsuccessful. The two short papers in the *Bulletin* and the longer one in the *Annals* were the only pieces he published on the subject. Neither did his fruitless occupation with Poincaré's Conjecture spawn any other work. In the twelve years between 1963 and his death he published only one more paper, an article dedicated to the memory of his erstwhile mentor, Ralph Fox.

But he never gave up hope. He even had plans to write a book about three-dimensional manifolds; the 160-page manuscript was found after his death. It contained a blank page entitled "Lemma 14," which was to be a crucial step in the proof of Poincaré's Conjecture. He was going to fill it in as soon as he found the way to do it. He never did.

The first person to seem to make progress was Elvira Strasser-Rapaport. She was the wife of the well-known Hungarian-born psychoanalyst David Rapaport, who had himself studied mathematics and physics before turning to studying the human mind. The two had met while staying on a kibbutz in Israel. Strasser-Rapaport had come late to mathematics, obtaining her Ph.D. only at age forty-three after she had raised the couple's two daughters. She did her doctoral work at New York University under Wilhelm Magnus, a former student of Max Dehn's. In 1964, while she was at the Polytechnic Institute in Brooklyn, Strasser-Rapaport did what Papa had hoped he would be able to do himself: She proved one of his conjectures. In her paper in the *Annals of Mathematics*, "Proof of a conjecture of Papakyriakopoulos," she uses techniques that she had learned from Magnus to prove the first of Papa's pair of conjectures. One down, one to go, she may have hoped…but then the efforts stalled.

Another attempt on Papa's program was undertaken in the early 1970s. About five years after Papa had given the reading list to Joan Birman, the deed bore some fruit, albeit in a slightly different direction from what Papa had dreamed about. Birman, who had obtained her Ph.D. from Magnus twelve years after Strasser-Rapaport, was then at Stevens Institute of Technology in New Jersey. She found a reduction of the Poincaré Conjecture to an algebraic problem different from Papa's. In the last section of her paper she states four remaining problems "whose solutions might lead to a resolution of the Poincaré Conjecture." Nobody seems to have attempted to pick up the challenge.

It is now known that Papa's approach is unworkable. In 1979 the Indian mathematician Anadaswarup Gadde (aka Gadde A. Swarup, aka Swarup Gadde) proposed a conjecture in the *Bulletin of the American Mathematical Society* and showed that it is implied by Papa's conjectures. Let's call Papa's conjectures P and Swarup's S. Swarup showed that P implies S. At this point, one must be careful: proving S would *not* prove P. On the other hand, disproving S *does* disprove P. This is known in mathematical logic as a contraposition. To illustrate, let us say that if you pack a lot (P), you need a large suitcase (S). From this it does not follow that if a traveler has a large suitcase, he actually packed a lot. But it does mean that if the suitcase is not large, he cannot have packed a lot. Hence, "not S" proves "not P."

Now, to disprove a proposition (not S) it suffices to find a counterexample. This is exactly what James McCool from the University of

Toronto did. His paper "A counterexample to conjectures by Papakyria-kopoulos and Swarup" was published in the *Proceedings of the American Mathematical Society* in 1981. So Papa's conjecture is definitely wrong (not P). In slightly over one page, McCool had dashed all remaining hopes that Papa's suggestions could eventually lead to a proof of Poin-caré's Conjecture.

But wait a minute, had Strasser-Rapaport not proved the conjecture seventeen years earlier? So who's right, Elvira or Jim? To resolve this question, recall that Papa had proposed not one, but two conjectures (P1 and P2), which in conjunction imply Poincaré's Conjecture. Swa-rup's paper showed that the pair of conjectures implies his conjecture S. When Jim provided a counterexample to S, it was clear that P1 and P2 in conjunction could not be true. Either P1 is wrong, or P2, or both. Since Elvira had proved that P1 is true, it was clear that P2 is wrong.

The long and the short of it is that, alas, Papa's efforts were for naught. Actually, the feeling among many mathematicians at the time was that the paper should never have been published in the prestigious *Annals of Mathematics*. Even though some parts of it were useful and could have been published elsewhere as a short paper, it altogether was certainly not of the standard that the *Annals* set for itself. According to one account, some of the *Annals'* editors even resigned in protest over the decision to publish the paper.

In the same year that Papa was born in Athens, Greece, halfway around the world a boy saw the light of day in Oakwood, Texas. His name was RH Bing. No, this is no typo, no periods are missing after the initials. The *R* and the *H* are not initials of his first names. They *are* his first name. He was called something like *Arhaitch*. The story is that Bing's father, a superintendent of a school in Texas and later a farmer, was called Rupert Henry. Naturally, he wanted his son to be named Rupert Henry Jr., but his wife thought that would sound too British for a Texas boy. So they settled on RH. Some older readers may remember JR, the eternal villain in the TV soap opera *Dallas*. Apparently double initials often moonlight as given names in Texas. Of course, the strange forename resulted in no end of confusion and stories. One of them goes that when Bing was named professor at the University of Wisconsin, he was asked what should be put on the nameplate. Loyal to his given name, he said, "*R* only, *H* only, Bing." When he arrived at his new office, he found the nameplate next to the door with the inscription "Ronly Honly Bing."

Bing's father died when the boy was only five, and his mother had to bring up her two children, RH and a little sister, on her own. Her job as a schoolteacher provided a meager income and the family lived frugally. She taught RH to read and do arithmetic long before he started primary school. Having jumped a class, he graduated from high school a year earlier than his contemporaries, Bing enrolled in Southwest Texas State Teachers College (later Texas State University), working in the cafeteria to make ends meet. Many years later he would be named the university's second Distinguished Alumnus, the first one having been Lyndon B. Johnson.

After obtaining his B.A., Bing became a high-school teacher of mathematics. At Palestine High School in Palestine, Texas, his responsibilities also included coaching the football team, training the track teams, and teaching the pupils how to type. In the hope of a salary increase, Bing started work toward an M.A. at the University of Texas at Austin. That is where he met Robert L. Moore, a well-known topologist with a reputation as an outstanding teacher. During his long career—he would teach until age eighty-six and even then was stopped only by order of the senate of his university—Moore supervised no less than fifty Ph.D. students. Bing was then in his midtwenties and thus a few years older than the average graduate student. This did not endear him to Moore, who did not like older students, especially those who had spent the previous years teaching high school. But Bing was in good company: Moore did not like black students either, nor Jewish students, nor female students, nor Yankees. A black student who wanted to join his class recounted that

Moore told him, "Okay, but you start with a grade C and can only go down from there." When he was faced with a discipline problem during his lectures, Moore is said to have on at least one occasion brought a six-shooter into class and laid it on the table before continuing to teach. The ex-high-school-teacher got into the professor's good graces, however, by proving himself a gifted mathematician, so much so that Moore eventually accepted him as a Ph.D. student. He even arranged a teaching position for him at the University of Texas, and during one class, Bing met his future wife, Mary Blanche Hobbs.

Moore thought that Bing's doctoral thesis—on obscure topological objects called planar webs—was outstanding, and it was published in the *Transactions of the American Mathematical Society*. Like all authors of learned papers, Bing received fifty free copies to distribute to interested colleagues. (This was before photocopying machines—and now e-mails—made such distributions superfluous.) But apparently nobody was interested, and shortly before his death Bing told friends that he still had forty-nine copies left.

Only a month after obtaining his doctorate, RH laid the foundations for his reputation by proving the so-called Kline Sphere Characterization Problem. The problem has a certain similarity to Poincaré's Conjecture: How does one recognize a two-dimensional sphere? It was already known that a one-dimensional object could be morphed into a circle if it could be separated into two pieces by removing two points from it. Bing gave a proof for the long-standing, but until then unproven, contention that a two-dimensional object could be morphed into a two-dimensional sphere, i.e., into the surface of a ball, on condition that it can be separated into two pieces by cutting along a circle, but not by simply cutting out two points.

Word of Bing's feat got around quickly, and he was promptly offered a position at Princeton University. There was one condition, though: He was to refrain from working in topology. One of the leading luminaries at Princeton at the time, Solomon Lefschetz, reportedly felt that the subject had little future. This seems rather astonishing given that Lefschetz himself had written the authoritative textbook on topology and was actively working in the field. Was the older man trying to stake off an exclusive domain for himself? This was not likely given established professors' penchant for collaborative efforts with bright young stars. Be that as it may, Bing declined the offer and in 1947 took a position at the University of Wisconsin instead. The most likely explanation is that Bing wanted to carve out a path of his own and not follow Lefschetz's lead. He stayed at Wisconsin for more than a quarter century.

Bing was a mathematician body and soul. A memorial tribute by the University of Texas at Austin tells the following story:

> It was a dark and stormy night when RH Bing volunteered to drive some stranded mathematicians from the fogged-in Madison airport to Chicago. Freezing rain pelted the windscreen and iced the roadway as Bing drove on—concentrating deeply on the mathematical theorem he was explaining. Soon the windshield was fogged from the energetic explanation. The passengers too had beaded brows, but their sweat arose from fear. As the mathematical description got brighter, the visibility got dimmer. Finally, the conferees felt a trace of hope for their survival when Bing reached forward—apparently to wipe off the moisture from the windshield. Their hope turned to horror when, instead, Bing drew a figure with his finger on the foggy pane and continued his proof—embellishing the illustration with arrows and helpful labels as needed for the demonstration.

In 1973 the University of Texas made him an offer of a professorship that he could not refuse. Bing became the highest-paid professor in the state of Texas, a far cry from the time when he had had to earn his living washing dishes and waiting on tables in the university cafeteria. His goal at the University of Texas was to raise the department's standard of research so that it would be ranked as one of the top ten state university mathematics departments in the country. He failed in that objective but made it to number fourteen, a very respectable achievement indeed. Bing and his wife were active in the Presbyterian Church, with RH serving as one of its elders. He was elected a member of the National Academy of Science, served as president of the Mathematical Association of America and president of the American Mathematical Society, was a member of the President's Committee on the National Medal of Science, and was showered with many honors. He died in 1986.

His topological contributions were numerous and often carried exotic names such as *pseudo-arcs, dogbone spaces, horned balls,* and *crumpled cubes.* One of his famous discoveries was the "house with two rooms." An object is said to be collapsible if it can be triangulated and then reduced by removing parts of it, one by one, until only a single point is left. An object is said to be contractible if it can be reduced to a single point by moving all its parts toward that point along certain paths inside the object. At first blush it would seem that these two definitions are descriptions of the same attribute, and indeed, every collapsible object is contractible. The opposite is not true, however, and Bing discovered a counterexample. It is the said house with two rooms and can be visualized as a cube with two floors. Entry to the top floor is through a tunnel from the bottom, and entry to the bottom floor is through a tunnel from the roof. Bing showed that the house with two rooms is contractible but noncollapsible.

Bing took to Poincaré's Conjecture like a fish to water. The fascination would last for most of his adult life. He never achieved his goal, obviously, because if he had, you would now be reading this book's final chapter. In fact, he was not even convinced that Poincaré's Conjecture was correct. "I suspect that perhaps the condition that loops can be shrunk to a point is not enough alone to insure that a three-dimensional manifold is topologically equal to the three-dimensional sphere," he wrote in one of his failed attempts to prove the conjecture.

Since doubts existed in Bing's mind as to the conjecture's veracity, he asked himself whether a different condition could be formulated that would be necessary, or sufficient, or both, to ensure that an object be recognized as a sphere. He sought a characteristic feature, sufficiently exclusive to ensure that all objects possessing it would be spheres. And he found one. It specified that all loops on the object under scrutiny must be able to move into small balls. He then proved rigorously that if the loops can be so contained, the object is, indeed, a sphere.

Let us explain this in a little more detail. Recall that Poincaré had conjectured that a manifold is a sphere if every loop on it can be shrunk to a point. Bing split the shrinking process into two parts. First, the loops are moved into small balls on the manifold. Then the loops are shrunk to a point. The second step is trivial: Once a loop is contained in the small ball, it can always be shrunk to a point. It is in the first step that Bing's approach differs from Poincaré's. While Poincaré's Conjecture considers all movements of loops without exception, in Bing's formulation the process of moving them into small balls must be performed without crossings, tanglings, or untanglings.

Toward the beginning of the year 1957, Bing submitted his work to the *Annals of Mathematics*. Like his earlier work on the Kline Sphere Characterization Problem, the resulting paper, "Necessary and Sufficient Conditions That a 3-Manifold Be S^3," falls under the general theme of recognition problems: How does one recognize a certain manifold, in this case a sphere, if one trips over it?

The theorem has nothing to say about manifolds whose loops need to be tangled to be moved to small balls. Hence it is weaker than what Poincaré—and Bing—had hoped for. But at least it was a proven theorem, not just a conjecture, and the editors duly accepted the paper. But then rumors surfaced that two attempts to prove the full Poincaré Conjecture were in the works at the Institute for Advanced Study in Princeton. Bing hesitated. His partial solution would look rather banal to readers in the future if shortly after its publication a full solution had been found. For a while he seriously considered withdrawing his paper. But then news came that the two attempts had "fallen flat" (Bing's words).

Bing's paper was published in July 1958. With it he had managed to pull off quite a feat, but, again, a proof of the Poincaré Conjecture it was not. Speaking of himself in the third person, the frustrated mathematician would write a few years later, "Bing made an attack on the Poincaré Conjecture…but achieved only a partial solution." It never ceased to surprise him that the Poincaré Conjecture remained unconquered, by him or by anyone else.

In the same year a voluminous paper appeared in the somewhat obscure *Mathematical Journal of Okayama University*. Its author was Ken'iti Koseki, who was also one of the journal's editors. The 107-page paper was entitled *"Poincarésche Vermutung in Topologie."* What's this, you may ask, a paper in a Japanese journal, by a Japanese author, written in German? Yes, all of Koseki's dozen or so articles in the journal were in German. How the refereeing was carried out is a mystery, but the final editing was certainly done by someone with a less-than-perfect knowledge of the German language—presumably Koseki himself. Like the title, the text, though understandable, is quite jolty.

Bing read the paper—most Americans who got their doctorates in mathematics at the time had to learn German—and declared that half of the "proof" restated some true result while the other half consisted of diffuse arguments. At one point, according to Bing, Koseki claimed without further justification that "if an obstruction arises, it would have been possible to start over." Such hand waving is certainly not acceptable in a mathematical proof, and Bing politely ends his overview of Koseki's proof by saying, "The argument is not clear." The knot theorist Hidetaka Terasaka from Osaka University was more straightforward. As a young man he had studied in Vienna, and he was therefore fluent in German. Charged by the *Mathematical Reviews* with an assessment of the purported proof, he did not mince his words: "It is surprising that such a primitive idea could lead to the confirmation of the Poincaré Conjecture." He continued, "Unfortunately the paper is ill compiled; it is hard to find out even where the Poincaré condition was essentially applied, and the details of this long proof are totally unintelligible to the reviewer." Quite deservedly, the paper was roundly ignored by the mathematics community.

Two years later Koseki tried to improve on his first attempt with another paper in the same journal, *"Bemerkung zu meiner Arbeit 'Poincarésche Vermutung'"* (Remarks on my paper "Poincaré's Conjecture"). Bing, and nearly everybody else, remained blissfully unaware of this follow-up. Only Terasaka weighed in again: "This is a very concrete exposition of the main idea underlying the author's previous proof of the Poincaré Conjecture; unfortunately, the author's main arguments are logically incorrect."

Nearly half a century later, Koseki's son Harutaka, himself a professor of mathematics at Mie University in Japan, wrote me a letter, saying that his father never really accepted that the proof was incorrect. As a young student Harutaka had talked to him about his proof, already in the knowledge that the mathematics community had not accepted it. He had the feeling that his father still believed in the basic value of his ideas but had lost the passion to try again.

Recall that Bing mentioned two attempts at Princeton to prove Poincaré's Conjecture that fell flat. He did not elaborate, but it is fairly certain that one of them was Papa's ill-fated effort. The other one was the attempt by Edwin E. Moise.

Moise was the person who proved in 1952 that the *Hauptvermutung* was correct for dimension three. After undergraduate studies at Tulane University he served in the U.S. Navy during the Second World War; he was a member of the group that broke the Japanese naval code. As a Ph.D. student at the University of Texas, he became Moore's student, one of the few Jews whom the master accepted as doctoral candidates. Moise would later say that he himself never felt any anti-Semitism at the university.

In the early 1950s, after completing his Ph.D., Moise was invited to the Institute for Advanced Study to serve as an assistant to one of the professors. He was sure of his capabilities…maybe a bit too sure. When he arrived, he announced that he was going to triangulate three-dimensional manifolds. And he did. Then he said he was going to prove the *Hauptvermutung*. And he did. Then he said he was going to prove Poincaré's Conjecture. And he did not.

Both Moise and Papa soon became aware that the other was working on the same problem. The competition as to who would finish first became quite intense. One day Papa announced—to Moise's consternation—that he had solved the problem and would present a seminar on his proof the following week. In due course a hole was found in the purported proof, and Moise could breathe a sigh of relief. A few weeks later, Moise announced a solution and presented it to the assembled colleagues. Of course, now a hole was found in his "proof," and it was Papa's turn to sigh a breath of relief. And so it went for a long while.

In short, Moise made no headway at all, though he spent a number of years on this pursuit. One of his students recounted that Moise's maxim was "When you run into a wall in your research, there are two approaches: Look for a way around it or keep hitting it with your head until either it or your head gives way." He used the latter approach. Eventually he seemed resigned that he would not be successful. "These results were developed in the hope that they might be used to produce a proof of the Poincaré Conjecture. Whether they can be so used, I do not know," he wrote in a summary of his work.

He never came to grips with his defeat. When it finally dawned upon him that he was not going to be the one to crack the conjecture, he turned away from mathematical research altogether. Instead, he dedicated himself to the improvement of mathematics education in American schools—a worthwhile subject, but not exactly the career path he had imagined. His textbook *Elementary Geometry from an Advanced Standpoint*—which was described by one critic as similar to "the seashore as viewed from a mountaintop"—was controversial from the start and ultimately abandoned.

Not to be put down, Moise moved on to literary criticism. He published six short notes on nineteenth-century English poets, but his contributions were not met with general enthusiasm. Apparently he was completely out of his depth. In one of his notes he explains that "La Belle Dame sans Merci" in John Keats's poem weeps because she was sexually frustrated when the knight did not respond to her moan. Writing from a relationship counselor's point of view, Moise thus managed to remove all of the beauty from the romantic poem. A professor of English literature described his article as amusing, but more suited for a psychology class than for an English course. After suffering two strokes, Moise died in December 1998.

Moise never knew that his 1952 *Annals* paper "Affine Structures on 3 Manifolds" would prove an essential ingredient of the ultimate proof of Poincaré's Conjecture. This crucial result, which was proven in another way by Bing in his 1959 *Annals* paper and given a streamlined proof by Peter Shalen in 1984, demonstrated that any three-manifold could be morphed into a shape consisting of a collection of glued-together tetrahedra. (This is called a triangulation.) An immediate consequence of this result is that all three-dimensional objects can be morphed into smooth shapes with no corners, thus allowing the problem to be studied using the field of analysis, an advanced form of calculus. Peter Shalen, from the University of Illinois at Chicago, writes, "You might ask whether the motivation for these papers involved opening the conjecture to a geometric analysis approach. I would say that, on the contrary, the motivation was to make it, and other topological problems, accessible to a combinatorial approach. In dimension three, the existence of a triangulation rather easily implies the existence of a smooth structure, and this leads to an entirely different approach—certainly not one that Moise or Bing or I were thinking about when we did our work on this particular problem." Yet solving the Poincaré Conjecture using geometric analysis was far in the future, and topologists and knot theorists had a lot more to say about the problem before the geometric analysts started to step in.

* * *

Mathematicians try to keep an open mind about things, and even though most of them believed that Poincaré could not have been wrong when he formulated his conjecture, some attempted to prove it wrong. After all, some of the best minds in mathematics—including Poincaré himself—had unsuccessfully tried to prove it. Maybe they had failed because the conjecture simply was not true? So some intrepid souls set out in this direction.

As I've noted, the simplest way to prove a hypothesis wrong is to provide a counterexample. The existence of a single counterexample is sufficient proof of a conjecture's falseness. This sounds easy enough: Simply build a three-dimensional manifold on whose surface all loops can be shrunk to points, but which cannot be squashed and squished into a sphere.

Bing was one of the mathematicians with an open mind. Without any indication as to whether the conjecture was true or not, he would alternately spend a couple of weeks trying to prove it, and a couple of weeks trying to construct a counterexample. During one of the latter two-week periods he suggested an attribute of knots he called Property P. This attribute has to do with certain cutting and sewing operations on manifolds known today as Dehn surgery—*Dehn* in honor of Max Dehn, whom we met in a previous chapter, and *surgery* because that is what cutting and sewing is all about.

The simplest way to perform Dehn surgery on a manifold, say a ball, is to first imagine a piece of knotted string inside this manifold. Remove the knot, including a neighborhood around it, from the manifold and sew a standard bagel into the cavity that has been created with a twist. You now have a somewhat warped object...not exactly the kind of beautified nose or cosmetically enhanced breast a Beverly Hills plastic surgeon would be particularly proud of, but then again Dehn surgery is judged by different standards. Anyway, many mathematicians went absolutely bananas about surgically altered manifolds, especially after Andrew H. Wallace and W. B. R. Lickorish independently proved a remarkable theorem about them in the early 1960s: All three-dimensional manifolds, except pathological ones similar to Möbius strips and Klein bottles, can be obtained from a three-dimensional sphere simply by performing a number of Dehn surgeries on it.

Back to Property P. A knot in a three-dimensional sphere is said to possess it if any manifold obtained by performing Dehn surgery on it is nonsimply connected. Hence, from a knot that does *not* have Property P, Dehn surgery would produce a simply connected object. The unknot, which can be untangled without cutting it, does not have Property P, but a manifold obtained by performing Dehn surgery on an unknot can easily be morphed back into a sphere through untwisting. To find a counterexample to the Poincaré Conjecture, one needs to find a nontrivial knot that doesn't have Property P: The resulting object would be simply connected but could not be morphed into sphere. And pop would go the Poincaré Conjecture.

For about half a century, Bing and others pursued their search for a counterexample by trying to find a nontrivial knot that does *not* fulfill Property P. None of them succeeded. By and by, the suspicion began to take shape that knots without Property P may not exist. Since most mathematicians believed the Poincaré Conjecture to be true anyway, this conviction soon became known as the Property P conjecture.

Then, in 2005, the reason all attempts to look for this particular counterexample had been doomed to failure became apparent. Nontrivial knots without Property P do not exist. Peter Kronheimer from Harvard University and Tomasz Mrowka from MIT asserted and proved this fact in the journal *Geometry and Topology*. Their paper was a real tour de force. The *Zentralblatt für Mathematik*, a journal that summarizes and comments on the tens of thousands of mathematical papers that are published every year, had this to say about Kronheimer and Mrowka's work: "This important paper...concludes the long history of the conjecture by the masterly combination of different approaches proposed over the years, together with different results from gauge theory and symplectic topology." In January 2007, Kronheimer and Mrowka were awarded the Veblen Prize by the American Mathematical Society "for their joint contributions to both three- and four-dimensional topology through the development of deep analytical techniques and applications."

"Masterly combination of different approaches" is the key phrase in the *Zentralblatt*'s review. The opening paragraph of the paper already makes one's head spin. The two authors state that they utilized as "ingredients" for their proof some recently discovered results: Taubes's theorem on the nonvanishing of the Seiberg-Witten invariants for symplectic four-manifolds, Gabai's theorem on the existence of taut foliations on three-manifolds with nonzero Betti numbers, Eliashberg and Thurston's construction of contact structures from a foliation, Floer's exact triangle for instanton Floer homology, another of Eliashberg's results on concave filling of contact three-manifolds, as well as a weaker version of an unproven conjecture of Ed Witten that relates the Seiberg-Witten invariants of smooth four-manifolds to those of Donaldson. (Ed Witten is the mathematical physicist who pioneered string theory. His research interests include topology, since string theory is closely linked, as one might expect, to the theory of knots.) After this introduction, the remainder of the paper is read at one's own risk.

Note that the fact that all knots possess Property P comes tantalizingly close, but does *not* prove Poincaré's Conjecture. Kronheimer and Mrowka's result says only that altering knots to spheres using Dehn surgery will not produce counterexamples. One would simply have to seek them elsewhere.

Wolfgang Haken was born in Germany in 1928. After the Second World War, he attended the University of Kiel, studying philosophy, physics, and mathematics and obtaining his doctorate in 1953. He then took a job as a researcher in microwave technology for the German industrial giant Siemens. While making his living in R&D, Haken kept up his interest in pure mathematical research. Working after hours, he found a mathematical technique to verify whether a tangled piece of string is knotted or not. With this he pioneered, in 1961, the algorithmic approach to topology. Electronic computers were in their infancy and engineers were excited about their possibilities. But the hopes reached far beyond engineering. Since Haken's step-by-step approach to the unknot problem could, in principle, be implemented on a computer, it was prophesied that the new toy would soon become applicable to pure mathematics. And so it would turn out to be.

Even though Haken's 130-page article was written and published in German, it made him well-known in mathematical circles, and he was asked to join the faculty of the University of Illinois at Urbana-Champaign as a visiting professor. The visiting position became a permanent position in 1965, when the university promoted him to full professor. Then his career really took off. Today he is best known for the solution of the then already more than hundred-year-old Four-Color Problem. The question was whether a geographical map with countries of any shape could be painted with four colors such that no adjoining countries will be painted with the same color.

At the University of Illinois, Haken teamed up with his four-years-younger colleague Kenneth Appel to devise a revolutionary new technique to find the answer. It became known as the brute-force technique and consisted in reducing the infinite number of coloring possibilities to a finite, if large, set of archetypical maps and letting a computer check each of them one by one. If a counterexample existed, it would have been found in this manner. Running hundreds of hours, Haken and Appel's computer found no map that required more than four colors. Since the two thousand archetypes that were checked represent all possible maps, the conclusion was that none existed. Not all mathematicians accept this number-crunching method as a valid mathematical proof. It cannot be checked in all its detail—it would take hundreds of man-years to replicate what the computer did in a few weeks—it lacks elegance, nothing is learned, no insights are gained. However, the proof, published as a book-length article in 1976, is considered valid, and computer proofs have become standard fare since then.

The proof of the Four-Color Theorem gained Haken worldwide recognition. What is less well-known is that before he set his sights on that famous problem, Haken had spent years trying to solve another famous problem. He was convinced he could prove Poincaré's Conjecture. Because his doctoral thesis had dealt with manifolds embedded in larger manifolds, he thought he was predestined to solve the problem, once and for all. After all, if all three-dimensional manifolds are just built of tetrahedra, then why can't one just check every possible combination of tetrahedra, using a computer? He become so obsessed with the problem that he was said to suffer from Poincaritis, an affliction that befell many a good mathematician in the twentieth century. It consisted in getting so wrapped up in trying to prove Poincaré's Conjecture that the sufferer would do little useful work for tens of years. Papa and Moise are believed to have been early sufferers of the then not-yet-diagnosed disease.

Another case study in Poincaritis is the erstwhile Romanian, now Frenchman, Valentin Poénaru. In the 1960s, traveling freely in Europe was not a matter of course. Eastern Bloc countries had to keep their borders closed to keep their citizens from leaving their beloved fatherlands. Many showed their "loyalty" by fleeing on foot or hidden in the trunks of cars over the borders. Athletes participating in international sporting events sometimes seized the occasion to defect to the West. Scientists who were, or pretended to be, loyal to their fatherland were allowed to travel to scientific conferences. Some of them also took off after their lectures. One such person was the Romanian mathematics student Valentin Poénaru. He was born in Bucharest and studied for his first degree at the university of that city. After a conference in Sweden in 1962, he didn't return to Romania. Already known for his mathematical abilities, he was granted asylum in France and obtained his Ph.D. in Paris in 1963. He spent the next four years at American universities, Harvard and Princeton among them. Back in France, he became professor of mathematics at the Université de Paris at Orsay, where he stayed until his retirement in 2001.

While visiting Princeton, Poénaru got bitten by the bug. Spurred by Papa's unrelenting attacks on the Poincaré Conjecture, Poénaru became interested in the problem. A competition ensued between him and Haken, a replay of sorts of the "Papa-Moise Show." Whenever one of them announced a proof, the other was quick to shoot it down and immediately publicize his own version. Then the procedure continued with the roles reversed. Over about ten years, their efforts never amounted to anything. With time, Haken began to get depressed, which is one of the side effects of Poincaritis. To his immense fortune, he found an antidote in the Four-Color Problem and got cured. The danger of relapse persisted for a long time afterward, however. He had given up hope of proving Poincaré's Conjecture but would declare from time to time that he was still looking...for a counterexample.

Unfortunately, Poénaru was too far gone to be cured. Thirty years later, in 1994, at a conference at Penn State University, he was still at it, outlining a program for the proof of Poincaré's Conjecture. It goes approximately like this. Take a simply connected three-dimensional body and extend it in an additional direction to make it four-dimensional. Add a handle so that it becomes a four-dimensional coffee cup. Poénaru proved that a handleless object could be squeezed, figuratively speaking, between the body and the coffee cup. Since the squeezed-in object is handleless, the smaller body must also be handleless. Now reduce the smaller, handleless body to its former three-dimensional state and show that it is a sphere.

The problem is that Poénaru never managed to prove all the steps in his program. Worse still, over the years it was shown that some parts of the program had to be wrong. Undeterred, Poénaru kept adapting his program and slogged on. And so it remained just that, a program. Ten years later, in 2004, he published a preprint on the Web site of the University of Paris in Orsay in which he is still seen describing his program to prove the Poincaré Conjecture.

John Stallings from UC Berkeley had been one of the first to identify the symptoms of Poincaritis. He had been an early sufferer before it had even been diagnosed as a mathematical disease. In 1966 he wrote, "I have committed the sin of falsely proving the Poincaré conjecture. But that was in another country; and besides, until now no one has known about it. Now, in the hope of deterring others from making similar mistakes, I shall describe my mistaken proof." He ends his mea culpa with some advice for budding mathematicians: "I was unable to find flaws in my 'proof' for quite a while, even though the error is very obvious. It was a psychological problem, a blindness, an excitement, an inhibition of reasoning by an underlying fear of being wrong. Techniques leading to the abandonment of such inhibitions should be cultivated by every honest mathematician." Would that this advice were also taken to heart by circle-squarers, angle-trisectors, and other quacks.

The efforts described in this chapter by no means encompass all attempts to prove or disprove Poincaré's Conjecture. We will have more to say about ill-fated endeavors in later parts of this book. But in the meantime let us turn from this rather depressing state of affairs to some more positive developments.

Chapter 9

Voyage to Higher Dimensions

The situation at the end of the 1960s was miserable: All efforts to prove Poincaré's Conjecture had been fruitless. Searches for a counterexample were equally unsuccessful. Mathematicians did not even know where to start. So they began looking for slightly different challenges in the wide void beyond the traditional Poincaré's Conjecture. They moved up to higher dimensions.

In one dimension Poincaré's Conjecture is trivial. Everybody knows a circle when one sees it. In two dimensions it is classical. Any two-dimensional surface on which loops can be contracted to a point must be transformable into the surface of a ball. The jump to three dimensions is—as we now know—quite formidable. Thus one may be forgiven for believing that the obstacles on the way to fourth, fifth, and higher dimensions should be even more difficult. Surprisingly, as it turned out, this is not the case. The present chapter is devoted to the search for answers to Poincaré's Conjecture in higher dimensions.

What exactly is Poincaré's Conjecture in higher dimensions? Recall the "first homotopy group," which we defined in chapter 7, using loops that are extended around a three-dimensional body. If all loops can be contracted to a point, the first homotopy group is said to be trivial. Poincaré's Conjecture claims that in this case the body can be morphed into a sphere. We also defined the second homotopy group using parachutes. Poincaré's Conjecture for four dimensions says that if all loops and all parachutes laid out on a four-dimensional body can be contracted to points, the body can be morphed into a four-dimensional sphere. Similarly, Poincaré's Conjecture for higher dimensions says that a body whose loops, parachutes, and higher-dimensional artifacts are contractible to points can be morphed into a sphere.

Progress did not advance in order of increasing dimension. Rather, within a two-year period, Poincaré's Conjecture was proved for every dimension from five upward. Only twenty years later was it also proved for four dimensions. It all started with a paper by Steve Smale. Or was it a paper by John Stallings? Or was it actually joint work by Stallings and Chris Zeeman? Did Raoul Bott help Smale fix up his proof? Or did he just spend a few relaxed days with him in the Swiss resort town of St.-Moritz? And did Andrew Wallace independently prove the theorem or did he only retrace Smale's work? Did I just write *independently* when I should actually have written *logically* independently?

Questions such as these are of extreme importance, if to nobody else, then at least to those involved. Who was first in proving what, when, and how, concerning the higher-dimensional Poincaré's Conjecture is a matter of no little bad feeling. Priority disputes in science are notorious, and mathematics is no exception. The "winner takes all" atmosphere in science does have its advantages, however. It makes people work hard and fast. Without pressure to be "first," science would advance at an unhurried pace, which would not be conducive to progress of human knowledge.

Thirty years after everything had been said and done, Steve Smale was still fuming. "I wish I could be more relaxed about this subject," he wrote in the *Mathematical Intelligencer*, then went at it once again, recounting in minute chronological detail events that had happened three decades earlier.

Smale was a dazzling maverick if ever there was one. Remember that a biographer of Henry Whitehead's wrote how unspectacular that mathematician's life was, "with little for the biographer to chronicle"?

Smale must be placed at the extreme other end in terms of life stories; there is nearly too much to chronicle. He was born in 1930 in Flint, Michigan. His family lived on a small farm outside town, and the boy attended the local school, where all nine grades were taught by one teacher in one room. Modest beginnings, we might say.

His father, a white-collar worker at a factory, was an avowed Marxist, not an easy position to take at a time when Senator Joseph McCarthy began to usher in the notorious era that would forever be associated with his name. At that time the boy was more interested in chemistry and amateur astronomy than in politics. In spite of the less-than-ideal conditions in the one-room school he attended, Smale did extremely well academically. In a statewide test that all kids took at the end of eighth grade, Steve scored highest out of about a thousand students.

Notwithstanding this success, Smale's teacher was skeptical about the boy's chances of attending college. He nevertheless applied and was accepted at the University of Michigan. Assisted by a small inheritance from his grandfather and a four-year tuition scholarship, he did not embark on an academic career but started building a track record of political activism. Smale became involved in left-wing politics, joined the Communist Party, demonstrated against the American involvement in the Korean War, visited Eastern Europe, attended a Communist youth festival in Berlin, and organized a Society for Peaceful Activities on his campus. So involved was he in political causes that he neglected his studies, continuing them only formally in order to avoid the draft.

Since Smale had so many extracurricular activities it is no surprise that in his senior year he was put on probation. Subsequently, he became more serious about his work at college. The student who had until then been a physics major switched to mathematics because it seemed to him an easier subject. In 1952 he received his B.S. and immediately applied for graduate school.

Smale was accepted but soon afterward got into trouble again because of his political activism. Eventually he was warned by the chairman of the math department that he would be kicked out if his grades did not improve. Faced with a choice of dropping out of either the university or the Communist Party, Smale wisely chose the latter course and henceforth devoted himself seriously to the study of mathematics. A year later the University of Michigan awarded him a master's degree and, in 1957, a doctorate. His reputation was tarnished, however. When he applied for his first job, the department chair wrote a mealymouthed letter of recommendation, pointing out that Smale was a "marginal, underachieving graduate student."

Clearly, this sentiment was not universal. In spite of his professors' misgivings Smale did stints at the University of Chicago, the Institute for Advanced Study in Princeton, the Collège de France in Paris, and Columbia University, and eventually embarked upon a tenure-track career at the University of California at Berkeley in 1964. He remained there for the next thirty years. Apart from doing research and teaching undergraduate and graduate courses, Smale supervised over forty Ph.D. students during his time at Berkeley.

But a leopard does not change his spots, and Smale was not about to abandon his outspoken ways. After all, it was the 1960s and where, if not at Berkeley, was one to speak one's mind? The McCarthy era had given way to the age of flower power, Woodstock, Haight-Ashbury. While the Vietnam War lurked ominously in the background, the young in California advocated free love, free speech, and free drugs. Smale teamed up with Jerry Rubin, the firebrand peace activist and cofounder of the Youth International Party (Yippies), in a movement to stop the war. Smale's acts of civil disobedience, such as trying to stop troop trains going through Berkeley, led to calls for his dismissal from the faculty.

While all this was going on, Smale also did some pretty good mathematics and began to establish himself as one of the world's leading topologists. Barely a year after obtaining his Ph.D., he astonished the world—at least the mathematical part of it—by showing that a sphere could be *everted*. This term refers to the turning inside out of a sphere if the skin is permitted to pass through itself, but no holes must be made and no ripping or creasing is allowed. Imagine an orange-colored basketball whose inside is colored black. Can this ball be turned inside out without ripping or creasing, such that after the procedure it is black on the outside and orange on the inside? Smale answered this question with a resounding yes. He proved a theorem that showed it is possible to evert the sphere. But the procedure implicit in his theorem was so complicated that nobody could visualize it. Smale, and others who tried, wanted to see with their own eyes how the sphere would evert. For once, the ability to visualize may actually have been a handicap. It was left to the blind mathematician Bernard Morin to devise a procedure to turn a sphere inside out that could actually be implemented (if the membrane was able to pass through itself, that is). By the way, can a one-dimensional sphere, a circle, be everted in two-dimensional space? No, it cannot. In the same manner that a circular train track cannot be switched from clockwise to counterclockwise without being lifted out of the plane, a circle cannot be twisted within the plane.

Throughout his entire career, which as of this writing is still going strong, Smale covered and even spawned a variety of fields. After the sphere eversion he did some pretty cool stuff with Poincaré's Conjecture, about which I will have more to tell below. Then he moved from topology to the theory of dynamical systems, then to mathematical economics, and recently to computer science. Every field he visited was significantly furthered by his achievements. In the introduction to a book that was published on the occasion of his sixtieth birthday, the authors noted, "Smale is one of the very few mathematicians whose work has opened up to research vast areas that were formerly inaccessible." They especially mentioned "his unusual ability to use creatively ideas from one subject in other, seemingly distant areas" and stressed that "in each case his innovative approach quickly became a standard research method."

One of his most notable contributions to dynamical systems was the "horseshoe" that was to become one of the icons of chaos theory. It neatly expressed the idea behind the *sensitive dependence on initial conditions*, which is one of the catchphrases of chaos theory—together with the *butterfly effect*, which, in principle, expresses the same thing. Imagine a dough in which raisins are embedded. The dough is folded like a horseshoe, then kneaded, stretched, and folded again, and kneaded, stretched, and folded again, and again, and again. During the stretching, raisins that initially lay close to one another sooner or later become separated, even if the initial distance between them was infinitely small. But since the dough occupies only a finite volume, the raisins cannot move infinitely far away from one another. Sooner or later they will become neighbors again due to folding.

In economics, Smale investigated whether market-clearing equilibria exist when buyers and sellers must agree on prices for goods and services, given the tastes of the former and the production schedules of the latter. Smale's intimate knowledge of dynamical systems then allowed him to investigate how prices adjust on their way to the equilibria. Turning to theoretical computer science, he used the skills he had honed during his interests both with dynamical systems and with economics to analyze how algorithms converge to their solutions. Along the way, Smale also provided important contributions to the physical and biological sciences.

In 1966, at age thirty-six, Smale was awarded the Fields Medal. As noted in the opening of this book, this is the most coveted distinction a mathematician can hope for. The award ceremony took place at the Congress of the International Mathematical Union in Moscow. On the flight to the capital of the Soviet Union, Paul Erdös, the legendary Hungarian mathematician, boarded the plane at an intermediate stop in Budapest. During the leg to Moscow, he told Smale about a rumor that a subpoena by the House Un-American Activities Committee awaited him back home. The rumor was true: Both Rubin and Smale had been subpoenaed, to the resentment of left-wing colleagues who suffered from "subpoena envy," as Rubin would later put it. Meanwhile Smale had the best of all worlds. He had the "honor" of being summoned by the committee without the hassle of having to be there.

The congress was as usual a grand affair, with thousands of mathematicians in attendance from all over the world. Smale would not let the occasion pass without some political commentary. On the stairs of Moscow University the prizewinner held an impromptu press conference at which he lambasted both the U.S.'s involvement in the Vietnam War and the Soviet Union's invasion of Hungary ten years earlier. As a consequence, he was briefly detained and interrogated by the Soviet authorities. The incident was widely reported in the American press and also in Soviet newspapers—sans the remarks about Hungary.

Some of his seminal work was done by the iconoclastic Smale in unlikely places. This would come to haunt him afterward. After he completed his Ph.D., the National Science Foundation accorded him a two-year postdoctoral fellowship. He spent the first year and a half at the Institute for Advanced Study at Princeton. For the last six months, he was invited to the Instituto de Matematica Pura e Aplicada (IMPA) in Rio de Janeiro. Smale and his wife packed a supply of diapers and traveled with their two young children to Brazil. In Rio he followed a strict schedule: Mornings were spent bodysurfing at the beach and evenings dancing the samba with the locals. During the afternoons he did discuss topology and differential equations with colleagues at IMPA.

But working only afternoons is not exactly what the NSF has in mind when it awards fellowships, and the administrators did not hesitate to make their feelings known to Smale. What made matters worse was that the flamboyant young professor was quite happy to trumpet that he often did his work in campgrounds, in hotels, on a steamship...or on ocean beaches. He did, however, also mention that he always had a pen and a pad of paper with him. And with this pen and pad he performed amazing feats. As he wrote in a letter of justification to the vice-chancellor of the University of California, some of his best-known work was done on the beaches of Rio. In fact, the horseshoe and the proof of the higher-dimensional Poincaré's Conjecture were conceived right there on Copacabana beach.

The NSF threatened to withhold its funds. After all, no government agency wants to be accused of paying for vacations and encouraging laziness. But what may be the correct approach toward lab scientists, perhaps microbiologists or psychologists, does not hold for mathematicians who have demonstrated again and again that they can do their work anywhere. Eventually Smale was able to convince the powers that be that results count, not the hours put in at an office.

For all his progressive, left-leaning politics, women's rights may not have been particularly high on Smale's list of priorities. In a notorious case, Jenny Harrison, an assistant professor of mathematics who had been denied tenure by Berkeley while men with possibly lesser credentials had been promoted, sued the University of California for gender discrimination. Actually, Smale had worked hard to bring women, including Harrison, into the Berkeley math department. But inexplicably, his attitude—at least toward her—shifted. She describes the events with sadness. She had been an up-and-coming star, "but suddenly everything changed. Smale encouraged me at first and told many he found my mathematical approach novel and exciting, but, in hindsight, his motives were not simple. I am not prepared to write about the unsettling experience that transpired as I prefer not to hurt anyone, no matter what they might have done to me." But she does mention that "Smale became inaccessible as a mathematician.... He would not give me an audience when I tried to show him my work. I tried several times to have a mathematical dialogue, but eventually gave up." Neither was teaching one of his stronger points. During classes, he often got the technical details mixed up, which prompted a distinguished colleague to remark that Smale "was nearly always locally wrong but globally right."

In 1998, as the twenty-first century approached, the International Mathematical Union (IMU) asked mathematicians to formulate a set of the most important mathematical questions for the coming century. The task was inspired by the list of twenty-three problems that the German mathematician David Hilbert had formulated in Paris at the congress of the IMU in 1900. Hilbert's problems had determined the direction of much of mathematical research during most of the first half of the twentieth century. Smale's list contained eighteen problems. He placed Poincaré's Conjecture as number 2, right behind the Riemann Conjecture, which had already been number 8 on Hilbert's list. Number 18 on Smale's list is "What are the limits to intelligence, both artificial and human?" In formulating this question, Smale was motivated by Gödel's incompleteness theorem, a set of ideas that limits the range of mathematical proof in a precise way. Smale took early retirement from Berkeley in 1994 to become Distinguished University Professor at City University of Hong Kong. He moved to the Toyota Technological Institute at Chicago, a privately funded institute for theoretical computer science, in 2002.

For his many achievements Smale was not only awarded the Fields Medal, but also the Veblen Prize, the National Medal of Science, the Chauvenet Prize, the von Neumann Award, and the Grand Cross of the Brasilian National Order of Scientific Merit. In May 2007 he was awarded the Wolf Prize in mathematics. The awards ceremony took place in Jerusalem, and the jury's description of Smale's work sounded nothing if not superlative. The laudation read,

> His proof in the early 1960s of the Poincaré Conjecture for dimensions bigger or equal to five is one of the great mathematical achievements of the twentieth century. Smale has reshaped the world's perception of dynamical systems. His theory of hyperbolic systems remains one of the major developments on the subject. Smale's work has contributed dramatically to change in the study of the topology and analysis of infinite-dimensional manifolds. In the 1970s, Smale's attention shifted to mechanics and economics, where he applied his ideas to topology and dynamics. The longest segment of his career, which took place in the early 1980s, focused on the theory of computation and computational mathematics. Against mainstream research on scientific computation, which centered on immediate solutions to

concrete problems, Smale developed a theory of continuous computation and complexity.

One rather unexpected distinction was a multipage profile about him in—of all places—the *Mineralogical Record*, a journal devoted to the collection of minerals. The reason for this unlikely journalistic piece is that Smale and his wife are avid collectors of natural crystals. The assemblage they have gathered over the decades is considered one of the world's finest private collections. A book with a hundred color plates of the couple's minerals was recently published under the title *The Smale Collection: Beauty in Natural Crystals.*

Smale's preoccupation with Poincaré's Conjecture began when he was still a doctoral student at the University of Michigan. He thought he had found a proof and excitedly rushed into a professor's office to tell him all about it. The professor did not say much while Smale outlined his proof. Only after leaving the office did he realize that he had not used any hypotheses on three-dimensional manifolds in his "proof." A few years later, while he was tanning on the beaches of Rio, Smale believed he had found a counterexample to the three-dimensional case. He wrote down the details, but when he reviewed his work, he found a mistake. So Smale's attack on Poincaré's Conjecture had got off to two false starts.

But he was not to be put off. His interests at that time had just started to center on dynamical systems, a theory that describes how manifolds move and flow when forces act upon them. So, while investigating how these topological objects develop over time, Smale hit upon a new way of decomposing them into cells. That, and his intimate knowledge of dynamical systems, would come in handy for a proof of the higher-dimensional Poincaré's Conjecture.

The crucial ingredient in Smale's proof was a new concept that he later developed into the so-called *h*-cobordism theorem. The word *cobordism* derives from the word *border* or *boundary* and refers to the connection between two manifolds. For example, if one manifold is a circle and the other manifold is a pair of disjoint circles, the cobordism would be something resembling a pair of pants: The single circle would be the waist and the pair of circles would be the exits for the legs. The preconditions of the theorem are that the cobordism fulfill certain requirements and that the two manifolds be simply connected, i.e., they must not be pretzels or similarly shaped objects. Given these postulates, Smale proved that the two manifolds are in some sense equivalent to each other—as long as everything takes place in dimension five or higher. A different interpretation of Smale's theorem is that the cobordism between the two manifolds is equivalent to an "open" cylinder—i.e., a pair of pants or an open-ended rubber pipe.

To prove Poincaré's Conjecture in higher dimensions one must ascertain that all contractible, compact manifolds can be morphed into a sphere (the surface of a ball). Let us investigate such a manifold. First, remove two disks from it. With the help of the h-cobordism theorem it can be shown that what remains of the manifold is equivalent to an open cylinder. Next, take the two disks and glue them together along their boundaries such that they form a ball. Then, by twisting and rearranging the manifold's parts, an equivalence relationship can be established between the cylinder and the two disks on the one hand, and the sphere formed by the two disks on the other. Hence, the original manifold under investigation is, indeed, topologically equivalent to the sphere, which is what Smale had set out to show. But beware: In the proof he had made use of the h-cobordism theorem, which is true only for dimensions five and greater. Hence, the version of Poincaré's Conjecture that Smale thus proved is also valid only for dimensions five and above. But Smale had done more than that. In his proof he had shown how certain manifolds could be constructed by gluing disks together. Thus he provided a starting point for the general classification of simply connected manifolds.

Smale mimeographed his proof in May 1960—it was the epoch before Xerox machines—and sent a copy to Samuel Eilenberg at Columbia University. Sammy, as everybody called him, was a Polish Jew who had escaped his fatherland just before the Nazis invaded it and was considered one of the world's leading topologists. He looked over Smale's proof, thought it was okay, and communicated it to the *Bulletin of the AMS*. There it was published as a short announcement.

At Oxford University in England, a freshly minted Princeton Ph.D. by the name of John Stallings heard a rumor that Smale had proved the higher-dimensional Poincaré's Conjecture. At first, he did not believe the news, but when Papakyriakopoulos confirmed that Eilenberg had vetted the proof, Stallings started to get interested. Not being aware of any details, he began looking for a proof himself.

Stallings, a mathematics student from Arkansas, had begun graduate school at Princeton in 1956. It was a time when traditions still commanded respect and students had to don black robes each evening for dinner. Stallings was a nonconformist, however, who did not quite take to customs and conventions. For example, a time-honored tradition at Princeton's math department was teatime in the common room. This daily routine gave, and still gives, wise professors and budding graduate students occasion to impress one another with clever talk about the latest gossip on the math circuit. One day, Stallings appeared at the event with a baby bottle, trying to look eccentric...or at least funny. He had filled it with hot chocolate, overlooking that a hot liquid requires an admixture of cool air to be drunk comfortably. Consequently the imprudent sucking on the baby bottle burned his mouth, and the episode did not turn out as funny as he had planned. In the further course of events, not disputed by Stallings, the baby bottle was filled with whisky. Some alcohol was always on the ready in Stallings's office. He needed it to get visiting speakers drunk before the start of their talks. This usually livened up the lectures, he maintained, and countered bouts of "attention surfeit syndrome" among the audience, a condition that describes the feigned attentiveness of listeners even when completely bored. (Reality television shows and pseudodocumentary films have recently been accused of using the same technique. Another example of mathematics on the cutting edge of culture, one might say.)

Stallings had arrived at Princeton at a portentous time when "topology was God and the Poincaré Conjecture was its prophet." A lifelong fascination ensued. Stallings, who would suffer bouts of Poincaritis throughout his life—as I already mentioned in chapter 8—accomplished great things, as we shall see below, but a proof of the conjecture in three dimensions eluded him. As can be surmised from the following tongue-in-cheek quote, he remained somewhat ambivalent about the matter: "When I die, I plan to be reborn in a universe with completely different laws of physics. I will be reborn as a counterexample to the Poincaré Conjecture. My colleagues and friends and mates will all be counterexamples to the Poincaré Conjecture. When two such counterexamples rub together, something wonderful will happen, something far more complicated than mere connected sum, to produce another completely different counterexample to the Poincaré Conjecture. But this too will be boring after a while. And so I shall finally prove the Poincaré Conjecture and destroy the universe!" Some people just never give up.

Early 1960 found Stallings at Oxford. He was still musing about the conjecture but most of the time he just stared at the blackboard in his office and drew blanks. The problems seemed insurmountable. But then news of Smale's breakthrough reached Oxford. The realization that a proof was possible removed the mental block from which Stallings had suffered. As he was staring at his blackboard some more, a path toward a proof suddenly occurred to him. It was totally different from Smale's, as it would later turn out. Omitting the details, let us just say that Stallings's proof consisted of stripping a manifold of dimension seven or higher to skeletons, embedding the skeletons in two balls, separating the skin from the inside of one of the balls, then recombining some of the items. And since a ball is a ball is a ball, Poincaré's Conjecture for all dimensions above seven was proven...again.

Meanwhile, on the other side of the globe, Smale was feeling pretty good. It was mid-June 1960, and he was still on the beach at Copacabana. More important, he had just proved the higher-dimensional Poincaré's

Conjecture. Now was the time to present his work to a wider audience in Europe. His first stop was Bonn, Germany, where he was hoping to give a lecture at the famous Mathematische Arbeitstagung, an annual affair, where the topics for presentation and discussion were decided upon on the spot by suggestions from the floor.

John Stallings was among the participants at the Arbeitstagung. He was a little downcast. He had just presented his version of the proof and had been told by a senior mathematician that he waved his hands too much—a euphemism that indicates a lack of rigor—at matters that had not properly been proved. Then it was Smale's turn to talk. To his horror, and Stallings's secret glee, a gap became apparent in the proof. Believing that this might be a fatal stab at his rival's proof, Stallings announced to the audience that there was a bug.

At first, Smale was devastated. But he soon collected himself and attempted to repair the gap. To his immense relief, the bug turned out to be squishable. Toiling feverishly for a week, he completed the task just in time for a topology conference in Zurich a week later. By the time he arrived in Switzerland, he had his act more or less cleaned up. Later he would claim that he had easily repaired the problem, but this does not seem to describe the events accurately. In fact, he also admitted that the week he spent in Bonn had been traumatic.

After the dramatic events, Smale took a few days off with his former Ph.D. adviser, Raoul Bott. In the Swiss resort town of St.-Moritz, the two walked through the Alps and may have had some further discussions about Poincaré's Conjecture. But even after all these efforts, the proof was not yet quite cleaned up. A friend who picked Smale up at the airport back in Rio, the mathematician Mauricio Peixoto, described him as haggard, tense, and tired. Obviously he was worried, and on the drive into town he admitted that there were objections to his proof, some of them serious. He was to spend the next three months cleaning and polishing a new version.

In the fall of 1960, Smale's proof was ready to go, and on October 11 the manuscript with the title "Generalized Poincaré Conjecture in

Dimensions Greater Than Four" was received by the *Annals of Mathematics*. Stallings's paper, on the other hand, had been communicated to the *Bulletin of the AMS* by Ed Moise in July 1960. Due to a fast production schedule, Stallings's article, "Polyhedral Homotopy-Spheres," which proved Poincaré's Conjecture for dimensions greater than seven, was published in the same year 112 pages behind Smale's first announcement.

The *Annals*' referees went over Smale's paper with a magnifying glass, and he had to revise his paper once again. The final version was sent to the editorial office in March 1961 and appeared in September 1961. Smale was fair: Stallings's work for dimensions greater or equal to seven, published a few months previously, was mentioned with full title and publishing details right there, on the first page of the paper. Now, this may seem just a wee bit more magnanimous than was absolutely necessary, and there was a subtle reason for this generosity. Once mention of Stallings's paper was made, nobody could accuse Smale of having ignored him. On the other hand, since mention had been made in the text, there was no need to list the paper again in the bibliography. The latter citation would have implied that Stallings's paper had been first and that Smale had perused it in his search for a proof. By omitting it from the references, a delicate hint was given to the cognoscenti. On the other hand, Stallings's paper made no mention of Smale's work at all, even though it must have been common knowledge that some article was forthcoming.

Stallings had used different methods from the ones Smale had used. Moreover, his proof held only for dimensions seven and up, while Smale's proof was valid for all dimensions greater than four. Could Stallings's method be extended to also cover dimensions five and six? Enter the English mathematician Christopher Zeeman, now known as Sir Christopher.

Zeeman was born in Japan in 1925, to a Danish father and an English mother. He was educated at a boarding school in England, which he compared to a prison in which he lost his self-esteem. Just how difficult life must have been for the young boy is apparent from his description of his subsequent service in the Royal Air Force (1943–47) as a breath of freedom. Zeeman was trained as a navigator for bombing action over Japan. A week before his unit was due to fly out, the Americans dropped the atomic bomb on Hiroshima and the war was all but over. Even though Zeeman was relieved to have been spared the task of bombing the land of his birth, he was somewhat disappointed at not seeing any real action. But it was just as well since the casualty rate among similar units was 60 percent. Later he would say that the atomic bomb had probably saved his life.

After his discharge from the Royal Air Force, Zeeman studied mathematics at Cambridge. Unfortunately he had forgotten most of his high school mathematics during his military stint and had to relearn much of it. But he did catch up and was awarded his Ph.D. in 1953. Teaching and research positions followed at the University of Chicago, Princeton University, Cambridge, and the Institut des Hautes Études Scientiques in France. In 1963 he was offered a new task. A friend of his took him to a muddy field at the edge of the town of Coventry in the West Midlands and told him two things: First, that this is where a completely new university—the University of Warwick—was to be established, and second, that he had been chosen to be the Founding Professor of Mathematics.

Zeeman had always thought his future to be at Cambridge, but when he was offered the opportunity to set up a department of mathematics, he found it difficult to stay loyal to his alma mater. It took a month of hesitation and a last sleepless night before he finally said yes. He never regretted it. With most universities established centuries ago, helping to build a new university from scratch was a challenge that does not often occur. It had the advantage that no strict hierarchy existed yet, decision making was more streamlined, and so everything could be created as one saw fit. Zeeman had largely a free hand, from the design of the building to the hiring of the faculty. He started by cajoling some leading lights to his soon-to-be-established department and eventually built it into one of the foremost mathematics institutes in the United Kingdom.

In 1991 he was knighted and became Sir Christopher. In 2005, on the fortieth anniversary of the university's establishment, and on his own eightieth birthday, the mathematics and statistics building was officially renamed the Zeeman Building.

Zeeman's best-known work, at least among nonmathematicians, is his contribution to catastrophe theory. Zeeman was partly to laud, or blame—depending on one's viewpoint—for the theory's foray into the mainstream...which ultimately led to its downfall. It was an early example of a mathematical theory in which the general public developed an interest. Cynics would say, however, that the application of catastrophe theory outside of mathematics was never more than a fad.

Catastrophe theory was developed by the French mathematician René Thom. It says, in a nutshell, that when no more than four variables are involved, discontinuities—sudden breaks in behavior due to small changes in the underlying variables—must occur in one of seven predefined ways. Zeeman visited Thom on a sabbatical in 1969–70 and learned all about the new theory. For Thom, whose interests lay halfway between mathematics and philosophy, catastrophe theory was exactly what it says it was: a theory. For Zeeman, situated halfway between mathematics and applied sciences, it was much more than that. It was a theory begging for applications.

At first, he devised the famous catastrophe machine—a simple contraption consisting of a rotating disk and a couple of elastic bands—which demonstrates beautifully how catastrophe theory works in physics. Then he applied the theory with some explanatory success to the buckling of beams, the collapse of bridges, the capsizing of ships. And then he himself went overboard. Zeeman and his followers started applying catastrophe theory to stock markets, linguistics, traffic flow, committee behavior, dogs' reaction to threats, and prison riots. Soon, every Tom, Dick, and Harry from softer sciences such as economics, sociology, and political science wanted in on the action. After a quick reading of Zeeman's papers or, worse still, of no more than some popular account of the theory, they were ready to give a pseudomathematical underpinning to hitherto unexplained phenomena in their field. All that time René Thom's serious and difficult treatise was one of the most frequently bought and least read books in science.

With time, the mathematics community became quite disillusioned with catastrophe theory. Smale played an important role in debunking it with a widely noted review of Zeeman's work in the *Bulletin of the AMS* in 1978. Thom later remarked that the criticism from Smale, someone he and Zeeman both admired, was one of their most painful experiences. (Neither Thom nor Zeeman held any grudge against Smale, as their tributes on the occasion of his sixtieth birthday testified.)

By and large, criticism was not directed against the mathematical underpinnings of Thom's work. Rather it centered on the indiscriminate use of the theory, also by Zeeman, for purported applications. Soon a backlash developed, and catastrophe theory, which had promised so much but produced so little outside of pure mathematics, sank into disrepute. Nowadays one hears little of it.

Zeeman spent the first year of his research career on Poincaré's Conjecture. He failed, but in an interview forty years later, he would recall it as one of his favorite failures. "A good mathematician probably has twenty-five failures to each success," he remarked in that interview. "The important thing is that new ideas keep coming." And they did. His preoccupation as a rookie mathematician with Poincaré's Conjecture provided him with the tools for understanding low-dimensional topology (where *low* means anything lower than six or seven).

The achievement that first brought him to international attention was a paper in the *Bulletin of the American Mathematical Society* in 1960. It was known that there are no knotted strings in four dimensions. (Recall that a piece of string whose ends are attached to each other is equivalent to a one-dimensional sphere.) This is actually amazing: Even the most tangled piece of string becomes an unknot in dimension four. Zeeman showed that the same result also holds for higher-dimensional knots. His paper was extremely short, even when compared to the terse style typical of a bulletin. In only eighteen lines he announced that two-dimensional spheres cannot be knotted in five-dimensional space. Two lines of the total were used to show that the result is quite general: any n-dimensional sphere is an unknot in a space of dimension higher than $3/2(n+1)$. So, any knot—regular or high-dimensional—magically unwinds itself as soon as it is transferred into a sufficiently high dimension.

A year later, in the same journal Zeeman published another, only slightly longer paper. In twenty-five lines with three bibliographic items (his previous paper had none), he announced that Stallings's proof of Poincaré's Conjecture in dimensions seven and higher could be extended to cover dimensions five and six. The announcement contained only the briefest of outlines and promised full details in a future paper that would carry the title "Isotopies of Manifolds." This paper never appeared. Instead, a more detailed account, with proofs, was given at a conference at the University of Georgia.

The germ for the conference had been planted in the spring of 1958 during a coffee-break conversation between the Texas topologist RH Bing and two colleagues. Actually, they decided to create a topology institute, but the term *institute* was used rather creatively. What the three men had in mind was some temporary gathering. The idea sounded good enough that the Office of Naval Research and the National Science Foundation agreed to fund the "Topology of 3-Manifolds Institute" at the University of Georgia during the academic year 1960–61. The program was then further streamlined to a four-week conference that eventually took place between August 14 and September 8, 1960. About forty topologists, professors and graduate students, showed up.

Apart from sending out invitations, the organizers had not burdened themselves with too much work. There had not even been any planning for the conference. Rather, an organizational meeting took place on the first day at which it was decided which subjects would be discussed. One of them was the Poincaré Conjecture. The talks and lectures that were presented at the "institute," or at least summaries, were published two years later in a volume entitled *Topology of 3-Manifolds and Related Topics*.

The book was dedicated to the English gentleman of this story, Henry Whitehead, who had planned to visit the University of Georgia, but died shortly beforehand.

Zeeman's paper starts out with a manifold—don't they all?—and then defines two sorts of exotic subspaces, which he calls trivial and inessential, even though they are anything but. He then goes on to explain how a subspace can be collapsed to another, smaller one. With these preliminaries out of the way, he proves two lemmas, minitheorems of sorts, that would later serve as stepping-stones in the proof of the real theorem. In the first, he shows that if a subspace can be collapsed to a trivial subspace, it must itself be trivial. In the second he says that every inessential subspace is wrapped up by a larger subspace, and this larger subspace can be collapsed to a smaller subspace.

The secret to extending Stallings's result to dimensions five and six consists of showing that a certain subspace of the manifold under inspection is trivial. Zeeman accomplished this task with the help of his two lemmas and a technique called *induction*. This is a method of proving theorems that goes back to the nineteenth-century Italian mathematician Giuseppe Peano. It consists of lining up mathematical objects and showing that if a certain property holds for one of them, it must also hold for its successor. And since every object has a successor, and the successor has a successor, the property must hold all the way up to infinity. So all one needs is to check whether the first mathematical statement is true, then the rest of the statements are true, one by one, like a row of falling dominoes.

Recall that the property that Zeeman had to prove was that a certain subspace of the investigated manifold be trivial. So he lined up the manifolds according to their dimensions and—using his two lemmas and the technique of induction—showed that if the subspace for one of them is trivial, the same is true in the next higher dimension. Hence the property that is required in order for the Poincaré Conjecture to be true holds from a certain dimension upward.

With this, the validity of the Poincaré Conjecture was pushed down from Stallings's seven dimensions to Zeeman's five. Why five? Why does the proof not also hold for dimensions three and four? Well, induction works all the way up to infinity, but not all the way down. It must start somewhere, otherwise it never gets going. The induction proof cannot start with any dimension lower than five because the subspaces in Zeeman's lemmas require sufficient elbow room. Only if the manifolds are large enough can the subspaces move around and collapse in a suitable fashion. Starting induction by knocking over the fifth domino, Zeeman proved Poincaré's Conjecture in dimensions five and up.

In the meantime, at Indiana University, Andrew Wallace tried his own wit and luck with the Poincaré Conjecture. He was well qualified, being one of the two codiscoverers of the important and famous Wallace-Lickorish Theorem. But he had got seriously stuck. In September 1960, frustrated, he wrote to Smale, telling him where exactly he was stuck and asking for details of his proof. Happy to oblige, Smale sent him preprints of his papers, and Wallace thanked him for them. Smale kept files of all his correspondence, including Wallace's acknowledgment, just in case anybody should ever doubt who had been first.

Upon studying the preprints, Wallace realized that up until he hit the snag, he had been working along similar lines as Smale. The helpful preprints provided the missing link and he could finally accomplish his mission. Obviously, his eventual proof retraced the path previously trodden by Smale. The basic question that needed to be answered was, as always, whether a certain manifold is topologically equivalent to a sphere. I will now outline Wallace's proof differently from the manner in which Smale's version was described above.

The first task consists of determining what components make up the manifold. From a topologist's point of view, even the most intricate manifold can be deemed to be a sphere with various, possibly knotted and jumbled handles attached to it. Examined by themselves, these handles can be straightened out and morphed into cylinders. The next task is to construct a look-alike of the manifold by building it up from the basic building blocks that have just been identified. So, starting with a sphere, the cylinders are jumbled and knotted in a suitable fashion and then attached to the sphere by surgery à la Dr. Frankenstein. Then corners, creases, and wrinkles are straightened out and flattened to make the Frankenfold smooth and even, until it is topologically equivalent to the original manifold.

To check whether the Frankenfold is equivalent to a sphere, the whole procedure is now retraced backward. The handles are rearranged, twisted, and canceled. Lo and behold, once the look-alike manifold has been reverse-engineered, one has arrived back at the sphere. This means that, for certain surgery processes, the original manifold is topologically equivalent to the sphere and thus the Poincaré Conjecture is proved. Unfortunately, in the course of all this surgery, the h-cobordism theorem had to be utilized, which, I remind the reader, holds only for dimensions greater than four. Hence the theorem is valid only for dimensions five and above.

With a little help from his friend, Wallace had done about as much as Smale had accomplished before him. To the public eye, however, it may have seemed that he had done it simultaneously. When he submitted his paper for publication, he did not choose one of the better-suited journals. Instead, he published it—under the cryptic title "Modifications and Cobounding Manifolds II"—in the *Journal of Mathematics and Mechanics*. Right after the introductory paragraphs, in which he describes his plan for the paper, Wallace states, "As a first application of these results, an affirmative answer is given to the Poincaré Conjecture for dimensions greater than five. The method is somewhat related to Smale's." That's putting it rather mildly. No wonder Smale was upset.

Meanwhile, across the Pacific Ocean, far removed from the mathematical hot spots of the time, an additional attempt to prove Poincaré's Conjecture in five dimensions took place. In splendid isolation and blissfully unaware of the works of Smale, Stallings, Zeeman, and Wallace, the Japanese mathematician Hiroshi Yamasuge published a paper in the *Journal of Mathematics of Osaka City University* in 1961, entitled "On Poincaré Conjecture for M^5." Yamasuge died young, at thirty-four, before his paper appeared in print, and history is rather silent on the author and this work.

On the occasion of the U.S. bicentennial celebrations in 1976, the American Mathematical Society invited the distinguished mathematician Paul Halmos to prepare an essay about recent advances in mathematics. It was a difficult work, written together with five colleagues, and one of the topics Halmos chose to describe was the Poincaré Conjecture in higher dimensions. At a meeting of the AMS in San Antonio he presented his paper, then "heaved several sighs of relief and went out to consume several beers—a little too soon."

Halmos had written, "The proof was obtained by Stallings for $n \geq 7$ (1960) and Zeeman for $n = 5$ and $n = 6$ (1961). At the same time Smale (1961), using completely different techniques, gave a proof for all $n \geq 5$." Halmos could not have picked a worse depiction of the events. Upon reading the manuscript, Smale shot off an indignant letter that was described by Halmos as strong, aggressive, belligerent, cantankerous, grouchy, and pugnacious. The letter referred to the "grossly distorted picture" that Halmos's talk had given and to the harm that it had done Smale. Halmos was fair and checked the record. Realizing that he had been incorrect, he took corrective action. The objectionable sentences were changed to "The proof was obtained by Smale (1960). Shortly thereafter, having heard of Smale's success, Stallings gave another proof for $n \geq 7$ (1960) and Zeeman extended it to $n = 5$ and $n = 6$ (1961)." While Wallace was roundly ignored, scrupulous justice had been done to Smale. But bad blood remained. If he had written a kinder letter, Halmos told Smale, he would have made the exact same changes and felt a lot less bitter about the whole affair.

It is just as well that what the Dutch mathematician Nicolaas H. Kuiper had to say a year later apparently did not come to Smale's attention. "S. Smale proved the Poincaré Conjecture for dimensions greater or equal to five." So far, nothing improper. But then Kuiper went on, putting between parentheses, "(For dimension equal to five, with the help of J. Stallings and E. C. Zeeman)." Accusing Smale of having needed help was like accusing the pope of requiring tutoring in Latin. And the parentheses did nothing to lessen the impact. If anything, they emphasized the accusation. Smale would really have blown his top.

But maybe he was right in complaining. Actually Stallings, Zeeman, and Wallace never claimed priority. But the fact that so many of Smale's colleagues got the chronology wrong shows that he had a valid point, even if it does not matter one bit either way to most of the world. Even ten years later, some writers still got it wrong. In 1988, John Milnor gave what Smale felt was an inaccurate account of the events: "The cases $n \geq 5$ were proved by Smale and independently by Stallings and Zeeman, and by Wallace, in 1960–61." Given the facts described so far, most readers would find no fault with this formulation, but, oh, no, Smale was not happy. In his view the word *independently* did not adequately describe what had actually happened. While willing to concede that "certainly Stallings and Zeeman, and Wallace, have done fine work on this subject"—does one hear just a teeny-weeny bit of condescension sounding through the praise?—he just could not keep himself from adding plaintively, "I do wish that mathematicians were more aware of the facts that I have just described." Willy-nilly, Milnor concurred, "What I should have said of course, is that the Stallings proof, completed by Zeeman for dimensions 5 and 6, was *logically* independent of your proof; but it is certainly true that yours came first. Similarly I should have said that Wallace's proof, for dimensions strictly greater than five, definitely came later, and was not really different from your proof." Now even Smale was satisfied.

Stallings was much more relaxed about the whole matter of priorities. In response to a letter by Zeeman, he wrote in 1988, "I got your letter to Milnor that must be about something he said or wrote about Smale, you and me in the 1960 era. I think I got some comment from

Smale about this too, but I cleaned up my office between then and now, and I have no idea what Smale said. And probably I should just go back to sleep."

That would have been a sensible thing to do, but Zeeman was too good a friend to be ignored, and so Stallings obliged with a minute account of his recollections. He ended it by saying, "I take these reminiscences of what happened 28 years ago with a little sense of amusement." Two weeks later he was even blunter: "If we had really been concerned about who thought of what first, we should have had a secretary take down all of our conversations.... It all sounds tiring and boring to me."

By 2003, the true course of events was well established. Picking up the threads in an account of the state of Poincaré's Conjecture after ninety-nine years, Milnor wrote, "Stephen Smale announced a proof of the Poincaré Conjecture in high dimensions in 1960. He was quickly followed by John Stallings, who used a completely different method, and by Andrew Wallace, who had been working along lines quite similar to those of Smale." Right, except that this time Zeeman was left standing in the cold.

The Poincaré Conjecture in dimension one is simple and not very interesting. Dimension two is more challenging but still manageable. Of the higher dimensions, seven and beyond are easiest to prove because such high-dimensional spaces provide sufficient elbow room. Conditions in dimensions five and six are more cramped and required more effort, but the proof could also be handled. So, within a year Smale, Stallings and Zeeman, and Wallace had proved the Poincaré Conjecture in dimensions five, six, seven, and everything beyond. The really interesting and challenging problems occur in dimensions three and four. Since most people had given up, for the time being, on three, all that remained was to prove Poincaré's Conjecture for four dimensions.

This task was left to Michael Freedman, an outstanding representative of a new generation of mathematicians. Freedman was about a quarter of a century younger than Smale, Stallings, Zeeman, and Wallace. Born in 1951 in Los Angeles, he entered Berkeley at age seventeen, but transferred to Princeton a year later. It took him barely four years to complete the requirements for all academic degrees, and at age twenty-two he was awarded a doctorate for his thesis in topology.

Moving back west, the young lecturer taught at Berkeley for two years—some of his students were older than he was—before moving east again for a year, to the Institute for Advanced Study at Princeton. When he reurned to the West, a professorial career at the University of California in San Diego followed, and he was quickly promoted through the ranks, ending up in 1985 with a named professorship. Along the way Freedman was appointed California Scientist of the Year, elected to the National Academy of Sciences, and named a MacArthur Fellow, all in 1984, elected to the American Academy of Arts and Sciences in 1985, awarded both the Fields Medal and the Veblen Prize in 1986, and presented with the National Medal of Science by President Ronald Reagan at the White House in 1987. And he was still only thirty-six years old.

In 1998, Freedman decided on a change of air. He widened his research interests to include computer science and moved, this time to the north, from San Diego to Redmond in the state of Washington. If Redmond is known for one thing, it is that Microsoft has its headquarters there, and that was exactly where Freedman was headed. He joined the theory group at Microsoft Research, being the first Fields Medalist ever to work in industry.

Microsoft Research's seven hundred employees do basic and applied research in computer science and software engineering, covering fifty-five research areas, such as speech recognition, information retrieval, graphics, machine learning, and many more. The theory group deals with alternative models of computation, combinatorics, graph theory, the theory of algorithms, and also with fundamental problems in theoretical computer science, such as whether P equals NP, or whether quantum computing will one day be possible.

Freedman found the environment at Microsoft Research enticing.

On the one hand, the academic model of professors working in ivory towers is balanced with the necessity of transferring the results to product-development teams. On the other hand, the theory group is permitted to operate outside the constraints of product cycles to concentrate on long-term visions. One possible impediment to joining the company was that the Microsoft Corporation offered its employees stock options instead of a pension plan. But that did not worry Freedman too much. An avid outdoorsman, he believed he was more likely to die in an avalanche than as a pensioner.

The Fields Medal was awarded to Freedman for his work on the classification problem, of which the Poincaré Conjecture is a part: Determine what it is that distinguishes spheres from other topological objects. But Freedman did more than just show how to recognize the four-dimensional sphere. He provided a complete classification of closed, simply-connected four-dimensional manifolds.

Four-dimensional space presents formidable problems. This is particularly galling because it is the space that we live in: three dimensions for length, height, and width and another for time. But neither Smale's and Wallace's nor Stallings's and Zeeman's methods function in four dimensions. In particular, the nice decomposition of manifolds into balls and handles, which Smale got to work so neatly for dimensions five and greater, breaks down in dimensions three and four.

One would believe that things are simple in low dimensions and become more complicated the higher up one goes. But the reverse may be true. To see how difficulties can arise when the dimension is too low, take a coin lying heads up on a tabletop. It can never be made to lie tails up just by moving it around the plane—unless, of course, the plane were twisted into a Möbius band. But the tabletop is required to be "oriented," which excludes Möbius bands. To make it lie tails up the coin has to be flipped over, and to perform such a move it must be lifted off the tabletop. Hence, operations can be performed in three dimensions that cannot be done in two dimensions.

The problem with manifolds in space is that two-dimensional surfaces can always be immersed but not always embedded in four dimensions. Let me define the terms and then explain. To embed a manifold means to place it into another, higher-dimensional manifold, without changing it topologically. The manifold and its image, sitting in the higher-dimensional manifold, become virtually indistinguishable. To immerse a manifold means to place it into another manifold, allowing intersections. An immersion looks like an embedding at every point, even though globally it is not.

Take, for example, the one-dimensional figure 0 as printed on this page. It is an embedding of the circle into the plane because, except for the printer's ink, it is indistinguishable from its surroundings. On the other hand, the figure 8—considered a twisted circle—is only an immersion of the circle into the page, not an embedding. Before it was immersed into the plane, the circle had to be twisted; therefore parts of the figure intersect on the printed page.

To accomplish his proof, Smale needed embeddings, not immersions. So he had to take something like the figure 8 and adjust it slightly to get an embedding. However, this necessitates the untwisting of the figure, and to do this it must be lifted into a third dimension. (Similarly a Klein bottle is only immersed into three-dimensional space, but with an extra dimension, it can be twisted slightly and become an embedding.) In fact with enough dimensions available, Smale was able to employ a technique called Whitney's trick, allowing him to twist things into embeddings. But without them, his cobordism theory fails. The upshot of this is that in four-dimensional space Dr. Frankenstein cannot perform surgery because he does not have sufficient room to move his patient around. Hence, handle theory and the h-cobordism theorem don't work without more dimensions.

Not until the mid-1970s did Andrew Casson, then at Cambridge University and later at the University of Texas at Austin, Berkeley, and Yale, adapt handle theory to four dimensions. A highly respected topologist who never bothered to actually get his Ph.D., Casson introduced "flexible" handles, which were later renamed Casson handles.

They are constructed as a union of "kinky" handles—handles that self-intersect and are flexible in the sense that they *can* be embedded in four-dimensional manifolds. Casson constructed the handles to mimic the proof of the *h*-cobordism theorem for four dimensions.

Freedman started his trek in 1978 by developing a technique to embed manifolds within embedded manifolds. The result, henceforth known as the reimbedding theorem, can be utilized to construct new Casson handles inside larger Casson handles. But the technique could not deal with all Casson handles. There remained a small collection of cell-like sets that resisted Freedman's explorations. Somewhat frustrated, he showed his work to a colleague, who immediately suggested shrinking the offending sets. Freedman did not make do with shrinking; he went all the way. Regarding the troublesome objects as "recalcitrant pockets of resistance," he "crushed them to points." Now, finally, progress was possible. Blasting the cell-like sets to smithereens had given Freedman a measure of control over the construction of Casson handles.

It took him another three years, however, to achieve his aim. Finally, in the fall of 1981, he was ready to show his work to the world. Freedman embarked on a road show of intensive seminars across the United States: a nine-day workshop in La Jolla, California, in August; a ten-day colloquium in Austin, Texas, in October; followed immediately by a one-week seminar at Princeton. Throughout these exhausting meetings the technical details of the paper were further developed and sharpened. Freedman thanked the participants not only for their interest and criticism, but also for their stamina "as the hours wore on."

The following year, his pathbreaking ninety-seven-page paper "The Topology of Four-Dimensional Manifolds" was published in the *Journal of Differential Geometry*. This paper earned him the Fields Medal. In a laudation, Jack Milnor warned of confusing simple-sounding results with simplicity of their proofs. "The simple nature of his results in the topological case must be contrasted with the extreme complications which are now known to occur in the study of differentiable and piecewise linear 4-manifolds." Milnor described Freedman's work as a tour de force: "His methods were so sharp as to actually provide a complete classification of all compact simply connected topological 4-manifolds, yielding many previously unknown examples of such manifolds, and many previously unknown homeomorphisms between known manifolds."

One of the implications of the paper is that the *Hauptvermutung*, discussed in chapter 8, is not universally true: A four-dimensional manifold may have different, and incompatible, triangulations. Another is the proof of the Poincaré Conjecture in four dimensions.

How so? Well, Freedman had managed to classify all closed four-dimensional manifolds with a trivial fundamental group, and the four-dimensional sphere is one of them. That is, Freedman could list all closed four-dimensional manifolds whose loops, or rubber bands as we called them before, can be stretched and shrunk to a point. Furthermore the only manifold whose parachutes also stretch and shrink to a point is the four-dimensional sphere. Poincaré had been vindicated also in dimension four.

Chapter 10

Inquisition—West Coast Style

The 1980s again saw a flurry of activity. At the time it was not at all clear whether Poincaré's Conjecture was true. Hence, many hopefuls were not trying to prove the conjecture but hoped to gain fame by finding a counterexample. One of the people prospecting for a counterexample was Steve Armentrout, from Eldorado, Texas.

Born in 1930, Armentrout entered the University of Texas in 1947. At commencement exercises four years later he was one of only four students, out of 371, to receive their bachelor's degree with highest honors. He then did his Ph.D. as yet another of R. L. Moore's disciples. RH Bing's junior, in terms of his doctorate, by eleven years, he later followed a career at the University of Iowa and then at Pennsylvania State University.

A few years after obtaining his Ph.D., Armentrout started suffering from a mild form of Poincaritis. He had became interested in the Poincaré Conjecture in the mid-1960s and stayed fascinated by it throughout his career. It was the underlying but often undeclared motivation for much of his work. At Penn State he started working on the subject in earnest. His plan of attack was to prove that a counterexample exists. He was not going to present it and he was not even going to show how it could be constructed. All he was going to do was to prove that it is constructible in a finite number of steps. He was going to provide a proof of concept, as it were. As tangible evidence of a counterexample's existence this may seem on the light side, but as a mathematical procedure the submission of an existence proof is quite acceptable. It would be admissible in a court of mathematics as conclusive evidence that the Poincaré Conjecture was wrong…even in the absence of the smoking gun.

If that sounds a wee bit complicated, watch what follows. Armentrout was going to accomplish his feat by indirect reasoning. Under the assumption that the Poincaré Conjecture is true, he was going to show that an object (which he called M) could be constructed that then leads to a contradiction. Using variations and elaborations of ideas of RH Bing, as well as results by Moise, he was going to prove that M both is, and is not, a three-dimensional sphere. Now an object that is a ball and at the same time is not a ball certainly is a contradiction if ever there was one. And if an assumption leads to a contradiction, then the assumption—in this case the Poincaré Conjecture—must be wrong.

However, there remained two slight difficulties. In Armentrout's words, "We have two main problems to consider. First, that of showing that M is a 3-sphere, and second, that of showing that M is not a 3-sphere." Well said. Other than that everything is simple.

Armentrout set out to devise a method that would allow the construction of a three-dimensional sphere with a special property. It would possess a triangulation that has no collapsible subdivision. This means that an obstruction prevents it from collapsing to a point. Hence, even though the constructed object would be a sphere in Poincaré's sense, it would not fulfill the requirement that loops on its surface can be shrunk to a point.

The eventual paper's title was clear and succinct: "A proof that the 3-dimensional Poincaré conjecture is false." The paper comprised six chapters and filled hundreds of typescript pages. In March 1981, Armentrout started sending copies of the first chapter to a number of colleagues. In the cover letter he explained that he was convinced of his paper's correctness but that he would nevertheless appreciate corrections, comments, criticisms, and suggestions. "It is particularly important to know whether there are mistakes that would invalidate the main result of the paper." Indeed.

After having sent out the first chapter, the others followed by and by. But then everything went rather quiet. It is not known whether one of Armentrout's colleagues found an error and told him about it in confidence, or whether nobody bothered checking the details. In any case, although ancillary results and spin-offs of his work did appear in print, Armentrout never published anything on the Poincaré Conjecture directly. Neither is the manuscript on the counterexample easily accessible. The only copy known to exist is in safekeeping in the archive of the American Institute of Mathematics in Palo Alto, California.

While we are on the subject, how can a prospector searching for the elusive counterexample ascertain that what he or she found is, in fact, a valid counterexample? One way would be to do it the old-fashioned way: prove it mathematically. The modern way would be to have a computer verify it. In 1994 the Australian mathematician Hyam Rubinstein presented an algorithm at the International Congress of Mathematicians in Zurich that he claimed would be able to determine whether a three-dimensional manifold is a sphere. If one has a three-dimensional object that one suspects may be a counterexample, all one has to do is feed it into the algorithm and wait for the yes or no answer. Rubinstein's program did not produce a counterexample, however, because no object whose loops contract to a point was ever fed into the program and shown not to be a sphere. But he did rigorously prove that the algorithm would work if ever such an object turned up. Obviously, the algorithm is no help at all in proving Poincaré's Conjecture. It could test more and more objects, but even if it never identified a counterexample, this would be no guarantee that one does not lie just around the corner. We will have more to say about Rubinstein's algorithm.

* * *

The next three endeavors do not represent direct attempts at solving the Poincaré Conjecture. Rather, they are searches for alternative ways to prove the elusive theorem. The two-stage strategy was the same in all cases: First, prove that Poincaré's Conjecture is equivalent to another, hopefully simpler, conjecture. Second, prove the easier conjecture. The attempts invariably broke down at the second stage.

One such venture was by Wlodimierz Jakobsche, a talented student from Warsaw, Poland. In 1965, a mathematical statement called the Bing-Borsuk theorem had been proven by...Bing and Borsuk. It dealt with certain spaces, let us call them *alpha-spaces*. Specifically, it said that all one- or two-dimensional alpha-spaces are manifolds. For three-dimensional alpha-spaces the statement was not proven and, like the Poincaré Conjecture, remained a conjecture. Quite unrelated, John Hempel had shown in 1976 that Poincaré's Conjecture is equivalent to the statement that something called fake three-cells cannot exist. Now Jakobsche was going to pull the two seemingly unconnected propositions together.

In a paper that appeared in 1980 in the Polish journal *Fundamenta Mathematicae*, he proved that if there does exist a fake three-cell, then at least one three-dimensional alpha-space must exist that is not a manifold. Now, if Bing and Borsuk's conjecture was true for three dimensions, then all alpha-spaces must be manifolds. This, in turn, would mean—by Jakobsche's theorem—that fake three-cells do not exist. And if no fake three-cells exist, then Poincaré's Conjecture is, by Hempel's theorem, true. So if only someone would prove the Bing-Borsuk conjecture for three dimensions, Poincaré's Conjecture would also be proven. Alas, nobody did. As for Jakobsche, after writing a few good papers, he went to the United States for a year and lost all further interest in mathematics.

In 1981, Thomas L. Thickstun, then at University College of North Wales in the United Kingdom, set out to replace the Poincaré Conjecture by another conjecture about manifolds that fulfill certain requirements. The proposed theorem reads, "Poincaré's Conjecture holds iff every open, irreducible, acyclic 3-manifold, which is a degree one proper image of an open 3-manifold embeddable in S^3, is also embeddable in S^3."

Don't be surprised by the double f in *iff*. It is math shorthand for "if and only if." The remainder of the statement may well put you off, however. So, to give a flavor of how mathematicians reason, here is another version of the statement, transposed to a fantasy setting. "Poincaré's Conjecture holds iff every glossy, small rectangular picture, which is a rough black-and-white photocopy of a glossy picture that can be stuck into an album, can itself be stuck into an album."

Why should Poincaré's Conjecture be true if this seemingly unrelated conjecture—the real one, not my analogy—holds? What is the relationship between the two? This is what Thickstun showed. In a two-page article in the March 1981 issue of the *Bulletin of the American Mathematical Society*, he outlined how he intended to demonstrate the equivalence of the two conjectures, promising full proofs elsewhere. And a good thing that was too because when the full proofs eventually appeared, three years and six years later, they ran to well over a hundred printed pages. About one of them a reviewer wrote, "The paper is skillfully written, providing ample detail for the devotee but also frequent scenic overviews for the less committed reader." One problem remained even afterward, however. Nobody managed to prove the new statement. Ultimately Thickstun's results shed no light on the Poincaré Conjecture. They did, however, constitute important progress on an unrelated problem in topology known as the Resolution Conjecture, which, in full generality, remains unproven to this day.

In 1983, David Gillman from UCLA and Dale Rolfsen from the University of British Columbia had a go at it. Remember Christopher Zeeman from the five- and six-dimensional version of Poincaré's conjecture? Well, this same Sir Christopher had proposed a conjecture way back in 1963 that said that every compact, two-dimensional polyhedron that can be shrunk to a single point by moving all of its points along certain paths inside the manifold can be thickened and then collapsed to a single point by triangulating it and then removing triangles, one by one, in an orderly fashion. This former method of shrinking is called contracting, the latter is called collapsing. He showed that this conjecture implies the Poincaré Conjecture. How? Well, start with a simply connected, compact three-dimensional body. Contract it to its two-dimensional "spine," that is, to its bare bones. Now, thicken the spine. By Zeeman's conjecture, it can be collapsed to a point. If this can truly be done—remember, it is only a conjecture—then the original body, after having been thickened to a four-dimensional manifold, can also be collapsed to a point. Now, if a four-dimensional manifold can be collapsed (by triangulating and removing triangles), it must have been—before shrinking—a four-dimensional ball. Hence, the original body must have been—before thickening—a three-dimensional ball. Confusing? Don't worry, the strategy of proving Poincaré's Conjecture by proving Zeeman's never worked anyway.

Because, unfortunately, nobody could prove Zeeman's conjecture. In fact, it was, and still is, widely believed to be too strong to be true. But twenty years later, in a paper published in the journal *Topology*, Gillman and Rolfsen showed that a weaker version of Zeeman's conjecture, let us call it "Zeeman Lite," is equivalent to Poincaré's Conjecture. So all it would take to prove Poincaré's Conjecture would be for someone to prove Zeeman Lite. Again, either nobody took up the gauntlet, or whoever did, did not get far. (To jump ahead for a moment, Perelman's eventual proof of the Poincaré Conjecture also provided proof for Gillman and Rolfsen's Zeeman Lite.)

On March 20, 1986, the English academic journal *Nature* carried a surprising announcement: The topologist Colin Rourke from the University of Warwick and his doctoral student Eduardo Rêgo from the University of Oporto had found a proof for the Poincaré Conjecture. The announcement was written by the well-known mathematician and math popularizer Ian Stewart, a colleague of Rourke's at Warwick. Since the general public does not generally read *Nature*, Stewart followed up with an article in the British newspaper *The Guardian*, and from there, the news item hopped around the world. For a while, Rourke and Rêgo could bask in the glory. Six months later, it was all over.

Rourke had received his bachelor's degree in 1963 and his Ph.D. three years later, both from Cambridge. After spending a couple of years at the Institute for Advanced Study in Princeton and at Queen Mary College in London, he settled down, in 1968, at the then just two-years-old University of Warwick. He was interested in the fundamental properties of space and started to make a name for himself as a topologist. In March 1985, the Portuguese student Eduardo Rêgo was writing a doctoral thesis under his supervision. In one of their discussions, Rêgo showed Rourke a theorem of his, and Rourke immediately thought that it would lead to a proof of the Poincaré Conjecture. Not wanting to be biased by history, they did not read up on previous attempts. By February 1986, they were convinced that they had a valid proof in their hands.

The community of mathematicians did not trust it. At the International Congress of Mathematicians in Berkeley, in August, there wasn't even any discussion of Rourke and Rêgo's work. Only three months after the congress had ended was Rourke invited to present his proof to the math department at Berkeley.

The American science magazine *Discover* would later describe the weeklong seminar at Berkeley as the "gentlemanly equivalent of an inquisition." True, there were no chains and no machines to inflict pain. But everybody seemed ready to torture Rourke. Some of the world's foremost topologists sat in the front seats of the seminar room, eager graduate students in the rows behind them. The audience just waited for Rourke to fall flat on his face. And eventually, he did.

Of course it was the job of those who attended to find possible holes in the mathematical argument. Rourke held out for four days, answering most questions and promising more details later for those questions he could not immediately answer. However, the listeners never seemed anything other than deeply skeptical, and the atmosphere bordered on the hostile. They just did not believe that a heretofore undistinguished mathematician could come up with something so momentous as a proof to the Poincaré Conjecture. On day five of the inquisition, they were proven right. An innocent question by a member of the audience opened a gaping hole. From this moment on, the proof began to crumble quickly. By the time the seminar was over, poor Rourke had to admit that there was a serious problem and that his manuscript was incomplete. It still is. In 2006, the CV on his Web site still lists a paper by himself and Rêgo entitled "A programme for a proof of the Poincaré Conjecture (advanced draft)."

Rourke returned to England. He now regretted all the publicity, especially the announcements to the press. His reputation temporarily in shatters, he had to start rebuilding his career nearly from scratch. Henceforth, his research results were received by colleagues with an additional grain of salt. Rourke has not lost his taste for grand questions, however, even if he has become more circumspect. In his latest endeavor—which lies somewhat outside his field of expertise—he proposed the hypothesis that a normal galaxy contains at its center a hypermassive black hole that generates the spiral arms. Admitting that his work was still preliminary and had obvious gaps, Rourke circulated his paper, "A new paradigm on the structure of the universe," without fanfare or press releases, "in the hope that others will help complete the work."

In mathematics, other than in biology, medicine, or physics, it is considered unseemly to make a run for the media. When the twenty-two-year-old Swedish student Elin Oxenhielm erroneously thought that she had solved Hilbert's sixteenth problem and wrote a press release about her feat in the fall of 2003, she publicly made a laughingstock of herself. The news was broadcast by the BBC, and Swedish newspapers touted her as a wonder girl. She gave interviews galore and dreamed of writing a book and even of making a movie about Hilbert's sixteenth, starring herself. Certainly a Fields Medal would just be a question of time. Instead, the bubble burst within days and her budding career was brought to a screeching halt before it had even begun.

Ian Stewart has no regret about having been instrumental in bring-ing the news of Rourke and Rêgo's purported proof to the public. He considers himself both a researcher and a journalist and is aware that to perform these tasks he must fulfill different requirements. When he wears his academic hat, Stewart makes sure that his research results make the traditional rounds of preprints, referee reports, revisions, and eventually, maybe two years later, publication in scientific journals. When he wears his journalist's hat, he looks for interesting stories for more immediate publication. A journalist cannot wait while the author circulates his preprints for comments, corrects the mistakes, sends the article off to a journal, waits for the referee to get around to reading it, and waits again while it goes through a lengthy publication process. "You must write the story as it is, warts and all, and await further devel-opments," Stewart once explained in an interview. And if the proof turns out to be wrong? Well, that will be another story then, won't it?

The Poincaré Conjecture never lost its allure. Every once in a while a new contender appeared. For example, He Bai-He from Jilin University in Changchun, China, announced in 1993, "A proof of 3-dimensional Poincaré conjecture." The announcement appeared, without any details, in the *Journal of Mathematical Research and Exposition*. This journal, published by the Institute of Applied Mathematics at Dalian University of Science Technology in the People's Republic of China, is—as may be expected of a publication whose articles are written mainly in Chi-nese—not widely read. Nothing was ever heard of Bai-He's proof again.

On May 9, 1999, Michelangelo Vaccaro of Rome popped onto the scene. On a Sunday evening he sent an e-mail around the world: "I, Mi-chele A. Vaccaro of Roma—Italy, with this message announce the fol-lowing proof of the Poincaré Conjecture. Title: 'Some characterizations of the 3-sphere.'" Big bark. In a few paragraphs, he outlined a purported proof and promised to send hard copies of his twenty-six-page paper to the postal addresses of everyone who was interested. Nothing else was heard of the proof, and Vaccaro disappeared. No bite.

Three years later, in April 2002, a hitherto rather obscure mathematician, Martin Dunwoody, from Southampton University in England, posted a paper on his university's Web site with the title "A proof of the Poincaré conjecture." This did create quite a stir. Again, newspapers and other media outlets showed interest. *The New York Times* carried the headline UK MATH WIZ MAY HAVE SOLVED PROBLEM and *Nature* exuberated BRITISH BRAIN AIMS TO GRAB MATHS MILLIONS. The Southampton University's public relations department was bombarded with phone calls. The paper's author, forewarned by the experiences of fellow travelers on the arduous path to Poincaré's Conjecture, refused to be interviewed.

Dunwoody had obtained his bachelor's degree at the University of Manchester in 1961 and his Ph.D. from Australian National University. After more than a quarter of a century at the University of Sussex, only interrupted by a year at Makerere University College in Uganda, Dunwoody received a call to the University of Southampton in 1992. Ten years later—Dunwoody was sixty-four years old and just before retirement—he posted his paper to the Internet. It would have been a brilliant ending to an otherwise somewhat undistinguished career.

Then the inquisition started. Dunwoody's strategy was inspired by Hyam Rubinstein's algorithm for recognizing the three-dimensional sphere. The paper was just six pages long and, posted on the Internet, was there for all to see. Of course, many did take a look…and found mistakes. Each time an error was pointed out to him, Dunwoody repaired it and posted a new version of his paper to the university's Web site. Some people objected to a mathematician posting no more than an outline of a proof, complete with mistakes, and expecting colleagues to point him in the right direction. Others disagreed, believing that there is nothing inherently wrong with a communal effort to further science. Anyway, Dunwoody was at version seven of his paper when disaster struck.

Colin Rourke had weighed in. Sixteen years earlier he had been in the exact same position as Dunwoody was now, and this was his opportunity to redeem himself. He knew exactly where to look for the weak points in the proof.

Within days, he had found it. It was on the paper's last page, where one would think that all that remained was to wrap up the proof. An unsupported statement said, "This 2-sphere will have the property that...the arcs...are uncrossed." An unsupported statement is the equivalent of the proverbial gaping hole. Dunwoody was at a loss. He tried hard to prove the assertion but did not succeed.

So he had to make some sad changes to version eight of his paper. First, he added a question mark to the title. It now read, "A proof of the Poincaré conjecture?" Second, he added a word to the first paragraph. Instead of "We give a proof of the Poincaré conjecture," it now read, "We give a prospective proof of the Poincaré conjecture." Third, and most important, he added a note in italics at the top of his paper: "Colin Rourke has pointed out there is a problem in the statement...," then added, "There may be an argument [to repair this]. (I hope!)"

His hope never materialized. On May 10, Hyam Rubinstein of the University of Melbourne, on whose work Dunwoody had built his proof, gave a seminar down under. "I believe I have an example showing a flaw in his approach which means a much deeper method is required." On May 15, a participant in an Internet discussion group ventured, "Dunwoody still tries to close the gaps. I expect that this will be hard work, perhaps it is impossible." And this is where Dunwoody's enterprise ended. No version nine, no proof of the Poincaré Conjecture.

Actually Colin Rourke had been laboring on his rehabilitation for quite a while before Dunwoody hit the limelight. He had continued work on Poincaré's Conjecture, but in a different direction. Since he had not succeeded in proving Poincaré's Conjecture, maybe he could...disprove it? Together with his student, he developed the Rourke-Rêgo-algorithm, which generates a complete list of manifolds that could, conceivably, be counterexamples. No input is needed for the RR-algorithm, it is a self-starter and then goes through the list. If a counter-example to Poincaré's Conjecture exists, the RR-algorithm will eventually find it.

But how can one know that a suspicious manifold is, indeed, a counterexample? As was pointed out above, Hyam Rubinstein had developed an algorithm, the R-algorithm, that checks whether a given manifold is topologically equivalent to the three-dimensional sphere. (It is the same algorithm that had inspired Dunwoody.) So whenever the RR-algorithm hits upon a manifold that is suspect, it is fed into the R-algorithm. Manifolds that cannot be deformed into spheres would immediately be flagged.

Combining the two algorithms into an RRR-algorithm, manifolds are produced and checked one by one. The RR-part produces manifolds and the R-part checks them. If Poincaré's Conjecture is false, the RRR-algorithm will come up with a counterexample in finite time. Rourke pulled all this together in a conference talk in Gokova, Turkey, in 1994. The article "Algorithms to disprove the Poincaré conjecture" was published in the *Turkish Journal of Mathematics* in 1997.

The problem is, of course, that if Poincaré's Conjecture is true, no counterexamples exist and the RRR-algorithm will churn through myriads of manifolds and run forever. One is reminded of the elusive search for the Higgs boson, the elementary particle that would explain the origins of the mass of other elementary particles. To confirm the boson's existence, the Large Hadron Collider is being built at the CERN facility on the Swiss-French border at a cost of more than $2 billion. The collider will be able to confirm the existence of Higgs bosons. But if such bosons do not exist, the collider will just keep running forever. What is certain is that running the RRR-algorithm to find counterexamples to Poincaré's Conjecture would be immensely cheaper than operating the Large Hadron Collider to find Higgs bosons. But then again, enormous costs are of little consequence to particle physicists.

* * *

On October 22, 2002, a twenty-one-page paper titled "Proof of the Poincaré conjecture" was posted to arXiv.org, an Internet repository for academic papers. In the weeks that followed, no less than seven different updates followed. By version number 8, posted on December 10, the paper had shrunk from twenty-one to six pages and the title had become "The Poincaré conjecture for stellar manifolds."

The author was Sergey Nikitin, a Russian mathematician who had obtained his master's degree from Moscow State University and, in 1987, his doctorate from the Academy of Science in Moscow. A year later, he was awarded the Lenin Komsomol Prize, a Soviet annual award for the best works in science, engineering, literature, or art carried out by an author not older than thirty-three. After four years at a German university, he has been at the University of Arizona since 1994.

This time the inquisition occurred wholly on the Internet. As soon as the first version was posted to the arXiv, a lively discussion ensued, with participants of Internet newsgroups expounding freely on the merits, or lack thereof, of Nikitin's paper. On October 31, Christophe Margerin from the École Polytechnique, Poincaré's alma mater, pointed out the first mistake. "Here is a copy of the e-mail I sent to Nikitin a couple of hours ago which gives a (rather simple-minded) counter-example to his main statement. The construction is very elementary and does not require any sophisticated concept nor argument, but it leaves almost no hope to fix Nikitin's 'proof.'"

Undeterred, Nikitin fixed and repaired. In the weeks following his first posting, he updated to version 2, then to versions 3, 4, 5, and 6. But errors remained. A month later, a newsgroup participant still took issue with the paper: "It is incorrect. His misuse of his induction hypothesis in the 'proof' of Theorem 2.1 is a critical mistake." The writer ends his posting by commenting, "I hope the readers of this newsgroup will excuse me if I do not take any more time to look over Mr. Nikitin's papers."

Nikitin was not to be put off. Version 7 followed, and a week later version 8, albeit preceded by some comments. "Theorem 2.8 was wrong in version 1 and that made the proof incomplete. It was fixed in version 2. The definition of simple connectivity was too strong in version 2. It was fixed in version 3. Version 4 has minor editorial changes that make presentation more clear and precise. Definition 3 is fixed in version 5. Versions 6, 7 and 8 contain improved presentations of the main result."

Is version 8 correct? It is doubtful, even though it has been sitting on the arXiv unchallenged since December 2002. But that is because nobody cares anymore. A new development had rendered the question all but moot. Just three weeks after Nikitin's first posting, in the period between his versions 5 and 6, another Russian mathematician had posted a paper to the arXiv. It would eclipse all previous attempts at proving the Poincaré Conjecture.

Chapter 11

Watching Things Go "Pop"

In the late 1970s and early 1980s William Thurston from the University of California, today at Cornell, formulated a spectacular hypothesis. He had been playing around with three dimensional manifolds, much as a child does with a set of LEGO bricks. After a while, he realized that every manifold is made of components of only a handful of basic shapes. While the LEGO company invents new shapes of bricks every now and then to keep customers happy (and to make them buy more bricks), nothing of the sort happens in geometry. Thurston proposed that the building blocks of which all three-dimensional manifolds are made up can take on only eight different geometries. Hence, any imaginable manifold must be built from these eight prototypes. Unfortunately, Thurston was not able to prove this; it was no more than a hunch. But an informed hunch by a first-class mathematician is more than just an inkling. It is a conjecture. Henceforth it was to become known as Thurston's Geometrization Conjecture.

Thurston was born in 1946 in Washington, D.C. He attended New College in Sarasota, Florida, and after receiving his bachelor's degree continued on to Berkeley. There his talents quickly became apparent. He completed his Ph.D. thesis in 1972 under Morris Hirsch and Steve Smale. His work was so profound that by 1974, at the age of twenty-eight, he was appointed full professor at Princeton University. Thurston was awarded the Veblen Prize in geometry and a Sloan Foundation Fellowship in 1974 for his work on foliations. Then he began to focus on low-dimensional topology again. His work in this area was described by a colleague not only as deep but strikingly original.

Not surprisingly, Thurston became a contender for the highest mathematical honor. At its meeting in the spring of 1982, the Fields Medal Committee decided to award a medal to Thurston. The official announcement would be made at the International Congress of Mathematicians that was to take place in Warsaw in the autumn of that year. However, political developments intervened.

Encouraged maybe by the election of Karol Wojtyla, aka John Paul II, as the head of the Catholic Church in 1978, the first Polish cardinal (and the first non-Italian since the sixteenth century) to be so honored, a new social movement, Solidarnosc (Solidarity), began to grow strong in Poland. With liberalism gaining ground, the Soviet Union felt its grip slipping, and on December 13, 1981, martial law was declared in Poland. General Jaruzelski became the new strongman. Many members of the ICM felt it inappropriate to hold a conference of such importance under these grim circumstances. In the hope that things would eventually calm down, the executive committee of the International Mathematical Union decided to postpone the congress for a year, to August 1983.

Things did calm down, the congress was held, and Thurston received his medal, albeit a year late. Because of the delay between the decision and the awards ceremony, the identities of the honorees—usually kept secret until the official announcement at the ICM's opening ceremony—were already known. The lack of uncertainty in no way diminished the recipients' achievements, however. The laudation extolled Thurston's fantastic geometric insight and vision: "His ideas have completely revolutionised the study of topology in two and three dimensions, and brought about a new and fruitful interplay between analysis, topology and geometry."

Exactly what were these revolutionary ideas? Thurston had set out to discover no less than the shapes of space. Before investigating how space is shaped, we must verify that all manifolds can be split into basic building blocks or, equivalently, that every manifold can be built up from such blocks. For this we require the notion of a prime manifold, or primafold, as I shall call it. Recall that prime numbers can be divided only by one and by themselves. Similarly, a primafold can be decomposed only into a manifold that is equivalent to a sphere and another that is equivalent to the original manifold. It cannot be split into more basic pieces.

Now that we know what primafolds are, how do they help us? In 1929, the German mathematician Adolf Kneser proved that every three-dimensional manifold can be decomposed into primafolds. So in a manner reminiscent of natural numbers—all of which can be formed by multiplying primes—manifolds can be created by combining ever more primafolds. Therefore they are the building blocks of which all manifolds are made. Kneser's theorem simplified three-dimensional topology enormously. It reduced all questions about manifolds to questions about primafolds.

How are primafolds combined to create more complicated manifolds? Beware, it is not simply a matter of gluing manifolds together like squashing wet bagels against each other. Recall the natural numbers. Large numbers are not created by just adding primes; they must be multiplied, which is a more complicated process. In topology, combining two manifolds is called surgery, and the name arose with good reason. Before a primafold can be combined with another, both must be opened up with a fine, if virtual scalpel. This means removing a patch of the primafold's surface. (In two dimensions the patch is a disk, in three dimensions it is a ball.) Only then can the boundaries of the two orifices be sewn together. All the while, care must be taken to preserve the directions of the orifices' boundaries. It's like a zipper: If it is to close, the zipper's two parts must be sewn into the fabric in corresponding directions. The removed patches—disks or balls—are henceforth ignored.

This can easily be visualized, but only in two dimensions. A bagel is a primafold: No matter where one takes a bite, what remains are the original bagel and the bite. The latter, being disklike, will be ignored. (I am speaking of a modest bite, not a gulp that could make the bagel into a cylinder.) Now suppose one were to take bites out of two bagels and zipper the bagles together at the openings that have thus arisen. One gets a figure eight that is thus not a primafold because it has been created out of two bagels. Adding another bagel, we get a pretzel, also not a primafold, and so on.

Two-dimensional manifolds were classified in 1907 by Poincaré and the German mathematician Paul Koebe when they, independently, proved the so-called uniformization theorem. The theorem implies that only three types of basic geometries exist in two dimensions. They are the Euclidean, the spherical, and the hyperbolical geometries.

Koebe's theorem can be used to completely describe the decomposition of surfaces into primafolds. This is done by geometry. In stark contrast to topologists, geometers don't just morph shapes at will. They are concerned with curvatures and regard, say, an ellipsoid as quite distinct from a sphere. Which shape is which depends on its curvature. The sphere, round in all directions, has positive curvature. The plane, being flat, obviously has zero curvature. Hyperbolic surfaces are shaped like saddles or mountain passes. Topologically they are equivalent to the flat plane, but geometrically they are not: They have negative curvature.

According to the uniformization theorem, these three basic geometries can be used to create all topological objects in two dimensions. Phrased differently, all objects must be created by a combination of these three basic geometries. The sphere, having no holes, is the surface of genus zero. The flat plane can be rolled up and zippered into a bagel or—with a little more effort—into a Klein bottle. Hence, these two objects are also considered flat. But since they possess a hole, they are said to have genus one. Finally, pretzels and other surfaces with genus two and higher can be created by cutting polygons out of hyperbolic space and zippering appropriate sides together. All two-dimensional surfaces with at least two holes—i.e., with genus two or higher—belong to the hyperbolic class of objects.

In three dimensions things become much more complicated. When Thurston took up his spectacular work in topology, all that was available to him was Koebe's uniformization theorem for two dimensions. He worked for years trying to understand the makeup of three-dimensional manifolds. Considering and rejecting various possibilities, he finally became convinced that he had found the Holy Grail: In three dimensions too, only finitely many geometries exist. Furthermore, he believed that he knew what their shapes were and announced his Geometrization Conjecture to the world. To recapitulate, Thurston speculated that the building blocks of which all three-dimensional manifolds are made, the primafolds, can be taken from a pool of no more than eight well-defined different geometries. By making use of shapes, Thurston, like Koebe before him, reintroduced geometry into topology by the back door.

So what are these eight geometries? What do they look like? As we saw above, the two-dimensional primafolds come in the guises of sphere, plane, and saddle. For three-dimensional manifolds the same shapes appear, albeit raised by a dimension. Hence we have the three-dimensional sphere (which is the surface of a four-dimensional ball), three-dimensional Euclidean space, and...well, whatever a saddle looks like when raised by an additional dimension. These primafolds are isotropic, which means that if you stand on one of them, the geometry is the same, no matter in which direction you look.

But there is more. Thurston conjectured that nature uses five additional geometries to build three-dimensional manifolds. The first two can be understood relatively easily. They are cylinders derived from one- and two-dimensional primafolds. Let me explain. The regular two-dimensional cylinder that we are familiar with can be thought of as being built up from two specimens of the one-dimensional primafold, the line. One line is bent into a circle, the other stands perpendicular to it. "Multiplying" the circle with the line, as it were, produces the cylinder.

In the same manner, three-dimensional cylinders can be built from lower-dimensional building blocks. Specifically, multiply a two-dimensional primafold—a sphere, a flat surface, or a saddle—with the one-dimensional primafold—the line—and you obtain a higher-dimensional cylinder. We disregard the product of the flat surface with a line, since it is just the previously mentioned Euclidean space. But a sphere multiplied by a line, and a saddle multiplied by a line, are definitely new building blocks. By the way, they are nonisotropic: In some directions one gets a curved view, while one direction shows a flat geometry. So these cylinders are two more primafolds.

Five down, three to go. But this is where we stop because the remaining building blocks are even more esoteric than the ones we already have. Two of them go by the names Nil and Sol, and the third doesn't even have a name. Suffice it to say that they are products, or twisted products, of two-dimensional building blocks.

Proving Thurston's conjecture is an even more ambitious undertaking than is providing proof for the Poincaré Conjecture. While the latter aims at identifying the manifolds that are equivalent to a sphere, Thurston's conjecture claims to classify *all* three-dimensional manifolds. Thurston managed to prove a diluted theorem, after making certain additional assumptions. He showed that his conjecture was correct for manifolds that are, in a certain sense, sufficiently large. Because of its difficulty and length, the proof came to be known as Thurston's Monster Theorem. As far as the general version of the conjecture is concerned—the one without the largeness assumption—it remained unproven.

Even though Thurston did not manage to prove the broad version of his conjecture, small advances were continually being made. And whenever something was proved, it was immediately added to Thurston's tally since he had been the first to conjecture it. With the "I told you so" attitude he had carved out for himself, a whole region of mathematics was henceforth considered his exclusive domain. Not everybody was happy with this. In a letter in 1988, John Stallings deplored "the crimes of Thurston" and of his colleagues. He explained that these misdeeds "consist of asserting things that are true if interpreted correctly, without really giving good proofs, thus claiming for themselves whole regions of mathematics and all the theorems therein, depriving the hard workers of well-earned credit." Nevertheless, Thurston had produced the structure that would ultimately lead to a proof of the Poincaré Conjecture.

Once the shell of a building has been constructed, one of the next steps is to decide how to heat the dwelling. The same holds when putting together primafolds to build a manifold. If Thurston may be considered the bricklayer, the heating technician came along in the person of Richard Hamilton.

Born in 1943 to a surgeon in Cincinnati, Hamilton received a head start in education through his parents and good teachers. He fortuitously developed an early and enduring love for three-dimensional geometry, which he ascribed—in response to the receipt of a prestigious mathematical award—to his high school geometry teacher, a Mrs. Becker. After attending Yale to study mathematics, graduating at twenty, he received his Ph.D. only three years later, in 1966, from Princeton University. As a young professor he showed a distinct affinity for the West Coast, teaching at the Irvine, San Diego, and Berkeley campuses of the University of California. He also held academic appointments at Cornell University, the Courant Institute at NYU, and the Institute for Advanced Study at Princeton. For a while he taught at the University of Hawaii, possibly because the surfing there was at least as good as in California. Hamilton, you see, is not the run-of-the-mill mathematics professor as popular imagination would picture him. He is flamboyant, loves horse riding and windsurfing, and at least one glossy magazine has described him as having a "steady string of girlfriends."

But life is not only about windsurfing and girlfriends. Hamilton is a serious mathematician and was awarded the Oswald Veblen Prize of the American Mathematical Society in 1996. He no longer chooses his home institution according to the surfing conditions and is now a professor of mathematics at Columbia University. The citation of the National Academy of Sciences, to which he was elected in 1999, spoke of Hamilton as the "creator and master of 'designer' evolution equations." Hamilton is to manifolds what Giorgio Armani is to men's suits.

His work, designing evolution equations, brought Hamilton to the attention of the famous mathematician Shing-Tung Yau. A co–Fields Medalist of Thurston's, Yau was cited at the International Congress of Mathematics in Warsaw in 1983 for his extremely deep and powerful work in differential geometry and in partial differential equations.

Yau's life story is the stuff of which novels are made. Born in 1949 in a farming community in the countryside of Hong Kong, Shing-Tung was one of three sons and five daughters. His father, a professor of Chinese literature and Western philosophy, took part in the establishment of the Chinese University in Hong Kong. The family was poor, food was scarce, there was no electricity and no running water. The children had to take their baths in a river.

When he was five years old, Shing-Tung took the entrance exam to a good public school but did not gain entrance because he failed mathematics. Instead he went to a small village school. There the kids were rough, and after first keeping his distance from them, Shing-Tung eventually became a leader of a gang of street kids.

Eventually he got into middle school. He started doing well, not because the teaching was so good, but because he listened attentively to the discussions his father had with his students. In high school he started studying geometry. The charm in proving elegant theorems based on simple axioms excited him. Then he turned to algebra and a serious love for mathematics started to develop.

To Shing-Tung's great chagrin, his father passed away when he was fourteen. The financial situation, not very good to start with, deteriorated. Only the strong will of his mother and some help by friends and former students of his father allowed Shing-Tung to continue his schooling.

Because he could not afford to buy books, Yau spent hours and hours in bookstores to read them there. When he was fifteen, he started earning money by tutoring younger students.

In 1966, he entered the Chinese University of Hong Kong. Even though he was very interested in history, and in spite of not having obtained the best grades in mathematics, he chose the latter as his subject. It was a fortuitous choice. "College mathematics opened my eyes. After I understood how mathematics works, I got so excited that I wrote a letter to my professor showing my great pleasure," Yau recounts.

From time to time, young American mathematicians from Berkeley and Princeton came to teach in Hong Kong. They were greatly impressed by Yau's abilities and helped him come to Berkeley, where he entered graduate school even though he had not completed college in Hong Kong. Thirsty for knowledge, he attended classes on all mathematical subjects from eight in the morning until five in the evening, sometimes eating lunch during a lecture. There he started proving some nontrivial problems, which brought him to the attention of the faculty. Some academic papers followed and the world-famous geometer Shiing-shen Chern, himself Chinese, agreed to become his doctoral adviser. In fact, Yau did not have to do much to obtain his Ph.D. Impressed by the student's work up till then, Chern decided that the papers Yau had already written fulfilled the requirements of a doctorate.

At Chern's advice, the freshly minted Ph.D. went to the Institute for Advanced Study at Princeton even though he had received an offer at twice the salary from Harvard. Positions at SUNY and Stanford followed, and Yau started solving some important problems. For example, he showed that the famous Calabi Conjecture was wrong, as everybody had expected. Only everybody had expected wrong, and soon Yau found a gap in his reasoning. It was a very painful period for him, for weeks he could not sleep. Then he started working in the right direction, eventually *proving* the Calabi Conjecture. It was to be instrumental in the development of string theory, which claims that space is made up of ten dimensions. The Calabi Conjecture says that the extra dimensions, beyond the four familiar ones of space-time, are tightly curled up, like the loops in a carpet. The resulting six-dimensional space is nowadays called the Calabi-Yau manifold.

Then he proved the Positive Mass Conjecture, which has significance for general relativity theory. Many more successes followed. Together with others, Yau established nonlinear analysis in geometry—which combines geometry, nonlinear analysis, algebraic geometry, mathematical physics—as a rich subject, essential for the understanding of nature. For his many contributions to mathematics, Yau received numerous honors. Apart from the Fields Medal, he was honored with the Oswald Veblen Prize in Geometry in 1981, was elected to the National Academy of Sciences in 1993, and received the Crafoord Prize of the Royal Swedish Academy of Sciences in 1994.

As we shall see in a later chapter, this Chinese mathematician was to play an important, if dubious, role in the drama of the ultimate solution to the Poincaré Conjecture. In any event, Yau and Hamilton struck up a friendship. What drew Yau to his colleague, apart from the latter's mathematical abilities, was that he enjoyed going swimming and having fun with him.

To study how to heat up—figuratively speaking—the manifolds built by Thurston, Hamilton introduced the concept of Ricci flow in the early 1980s. This is a differential equation, especially designed by Hamilton, that mimics the flow of heat through a body. Named after the nineteenth-century mathematician Gregorio Ricci-Curbastro, this flow plays a central role in the search for a proof of the Poincaré Conjecture.

Yau immediately recognized that Hamilton was onto something. A quarter of a centry later, Hamilton remembers that Yau "pointed out to me way back then that the Ricci flow...could lead to a proof of the Poincaré conjecture." In particular, Yau noticed that the Ricci flow might break a manifold into its primafolds. Of course this was only a suggestion, and it would need significant work to pull it off. But convinced that this would be the way to prove not only Poincaré's but also Thurston's conjecture, Yau encouraged Hamilton to continue on his path.

Bet let us begin at the beginning. Why did Hamilton name the flow after Ricci? Ricci-Curbastro was born in 1853 in Lugo, a town in today's Italy that was then part of the Papal States. His father was a well-known engineer, and the family enjoyed considerable social status. Gregorio and his brother did not attend schools but were educated by tutors at home. When he entered the University of Rome, at age sixteen, he was well prepared. But political turmoil ensued and Ricci-Curbastro returned home after barely a year. Subsequently he attended the University of Bologna and then Pisa. There he befriended Enrico Betti, who became his thesis adviser. Ricci-Curbastro impressed his teachers with his abilities. In 1877 a scholarship allowed him to spend a year abroad, and he chose Germany to attend the Technische Hochschule in Munich. This is where Felix Klein, the towering mathematical intellect of the time, discoverer of the fiendish Klein bottles, and a man greatly admired by Ricci-Curbastro, was active. The admiration was reciprocated with Klein holding the student from Italy in great esteem. Upon his return to Pisa, Ricci-Curbastro first became an assistant at the university and was eventually named professor of mathematical physics. He remained in this post until his death in 1925.

Ricci-Curbastro's most lasting contribution—the tensor—was introduced in a paper that he wrote together with his student Tullio Levi-Civita in 1900. The tensor is a generalization of the notion of scalars, vectors, and matrices. It is an array of numbers that describe physical or geometric quantities. A tensor can be of rank zero, in which case it is just a number that describes, say, a body's mass or temperature. It could be of rank one, in which case it is a vector that describes, for example, the forces acting upon a body. A tensor of rank two is a matrix and could describe, for example, the stresses emanating from three directions onto a three-dimensional body. In geometry, tensors are used to describe how a space bends differently in different directions.

For a long time the paper of the two Italian mathematicians remained on the sidelines, largely unnoticed by the mathematical community. Whoever did notice it dismissed it as no more than a technical accomplishment. But a dozen years after its publication, Albert Einstein began utilizing tensors when he reduced gravitation to a geometrical phenomenon. From then on, Ricci-Curbastro's contribution was recognized as the profound innovation that it really was. In fact, when moving beyond special relativity, Einstein found Ricci-Curbastro's methods so important that the mathematical theory of general relativity is formulated entirely in the language of tensors. Actually, Einstein, whose mathematical abilities were not quite as top-notch as his contributions to physics, had not found out about tensors by himself. Rather, it was either his friend the geometer Marcel Grossmann or Levi-Civita who introduced Einstein—with some difficulty—to tensor calculus.

The notion of the tensor leads to the idea of the Ricci flow. As pointed out above, the Ricci flow is a differential equation that is inspired by thermodynamics. Put a candle under a hot plate and heat starts dispersing. After a while, a uniform temperature prevails throughout the hot plate. Fire up the burner in the basement, and a little later the whole house becomes warm. A freezing mountain climber takes a drink—tea or vodka—and soon his innards get warm. The flow and the distribution of heat throughout the hot plate, the house, or the human body are described by a partial differential equation that was introduced by Joseph Fourier in 1822, in his treatise *Théorie analytique de la chaleur*.

The Ricci flow does something similar. Only this time it is not heat that is being dispersed throughout, but curvature. This may sound confusing—how can a geometric attribute be dispersed?—so let me start by elaborating on the notion of curvature.

One can easily picture what is meant by the curvature of a line winding along a plane. The tighter the line bends, the greater is its curvature, and vice versa. A straight line, for example, has zero curvature. At every point along the line the curvature is given by a single number, specifically the inverse of the radius of the osculating circle, i.e., the circle that best fits the curve at this point. The tighter the curve, the smaller the radius, the larger the curvature. However, for lines that do not just wind along a plane but snake through space, one number does not suffice.

The lines can bend and curve in more than one direction. To fully account for a line's curvature in space, two numbers are needed, one describing the curvature in the x-direction, the second in the z-direction. The same is true for surfaces floating in space. A cylinder, for example, is curved in one direction, but is flat—that is, has zero curvature—in the other. Hence, the sharpness of the bends in all directions must be taken into account when describing curvature in three- or higher-dimensional space.

Curvature of three-dimensional manifolds suspended in higher-dimensional space is more complicated still. One cannot express it simply as a couple of numbers. To describe the curvature of such a manifold, a whole array of numbers is required for every single point on the body. This is done with the tool that Ricci-Curbastro and Levi-Civita introduced into mathematics, the tensor.

Now let us imagine a manifold that changes its shape over time, like a jellyfish floating majestically through the water. But while the jellyfish's shape changes more or less cyclically, we want the manifold's shape to evolve toward a certain form. And while Joseph Fourier's heat equation takes its cue from nature, describing how temperature changes at each point of the body, the Ricci flow, as intelligently designed by Hamilton, *dictates* the evolution of the body's shape. Hamilton wanted to mold manifolds, to warp them in a way that would even things out, just as heat distributes across a room. And in his first paper on the Ricci flow he did exactly that, showing that certain special three-dimensional manifolds bend themselves into spheres.

It was not intelligent design versus evolution, but intelligent design *for* evolution. Hamilton's intelligently designed Ricci flow was to make manifolds evolve purposefully toward certain shapes. The Great Designer, in the human form of Richard Hamilton, had devised a differential equation that was going to morph even the most intricate manifolds into ones of simpler shapes. Incidentally, the sci-fi author Tina S. Chang has written a story in which gods use Ricci flow to manipulate the universe. Mad scientists indeed.

As envisaged by Richard Hamilton, the Ricci flow does to manifolds what Botox injections do to aging movie stars: They make frown lines, forehead creases, and crow's-feet disappear and render the lady's or the gentleman's skin smooth and fresh, as it was when she or he was young. And they do so in a delicate fashion without any injuries. The lady is the same lady, and the gentleman is the same gentleman, both before and after the injections; only their appearances have changed. Since the injections cause no injury, Botox obtained approval by the FDA and has been enthusiastically received by professional beauticians, who successfully—and quite profitably—administer the injections to millions of clients.

The situation is similar for Ricci flows. Well...sort of. Ricci flows are, indeed, designed to iron out wrinkles, warps, and creases from manifolds. And they do so without tearing or ripping. The manifolds are the same manifolds, both before and after the treatment, because only their geometric appearances have changed. Hence Ricci flows were sanctioned and enthusiastically received by professional topologists. (Alas, no money has yet been made from administering Ricci flows, even though, as will be told in chapter 14, a million dollars is waiting.)

We won't go into how the active ingredient of Botox, the botulinum toxin, blocks nerve impulses and temporarily paralyzes the muscles that cause the wrinkles. But here is a bit more about how the Ricci flow functions. The differential equation—that is, the Ricci flow—relates the Ricci tensor to the evolution of the manifold's scale. The former, you may recall, incorporates the manifold's curvatures, while the latter is a measure for distances on the manifold. Significantly, the Ricci flow relates the two quantities at every point by a minus sign. This means that the manifold expands wherever it is negatively curved and contracts wherever it is positively curved. This continues until all dents and bulges have been smoothed out. And then what happens?

Let us assume that the manifold is now positively curved throughout, even though some regions may be more warped than others. The negative relationship between curvature and scale implies that the more curved a manifold's region, the smaller the scale becomes. Now, when scale decreases, close points are pulled together, the result being that curvature increases. The upshot of this is that a positively curved manifold will curl up more and more, until it becomes a rapidly shrinking manifold of constant curvature. So it curls up tighter and tighter, becoming smaller and smaller. After a while the by now teeny-weeny sphere cannot be reduced anymore. At that point it just goes "pop" and disappears into thin air.

Thus, Hamilton's idea was simple enough. Take any beaten-up, kinked, and twisted manifold, let the Ricci flow do its thing, and watch while the manifold tries to bend itself into shape. Into which shape will it develop? If it eventually evolves toward one of the eight primafolds, or to a combination thereof, Thurston's conjecture is correct. And now comes the icing on the cake: If every beaten-up, kinked, and twisted but *simply connected* manifold ends up going "pop," Poincaré's Conjecture is proven!

The Ricci flow was designed to make visible the shape of even the most intricately warped and twisted manifold by unwarping and untwisting it until its shape is so simple that it can be recognized. The method is reminiscent of the plastic castles that can be found at amusement parks for children. In the morning, before the park opens its gates, the plastic canopy lies in a messy heap on the ground and one has no idea what shape it is. Only when it is being filled with air can you start telling into what kind of castle shape it turns.

But who has the time to sit around and watch manifolds being blown up? And even if someone did have time on his hands, it is not possible to study individually all manifolds, no matter how much time you have. The only alternative to this impossible task is to rigorously prove that a manifold—any manifold—must evolve toward one or more of Thurston's basic building blocks, and that the simply connected ones among them end up going "pop."

Hamilton started thinking about all this in 1979. Three years later, in

1982, he introduced the mathematical world to the Ricci flow. His paper entitled "Three-manifolds with positive Ricci curvature" was published in the *Journal of Differential Geometry* and made a big splash. In it he proved that under the action of the Ricci flow, manifolds with positive, but differing, curvatures evolve toward manifolds whose curvatures are constant in every direction. Thus the first step was taken to proving the elusive Poincaré Conjecture. By suggesting a proof strategy based on the Ricci flow, Hamilton had done a remarkable thing: He had proposed to solve a problem in one discipline (topology) with the tools of another (differential equations).

Encouraged by Yau, Hamilton set himself the task of showing which situations could occur when Ricci flows are allowed to operate on any kind of manifold. His hope was, of course, that all of them would end up becoming one or more of Thurston's building blocks, and that the simply connected ones would eventually go "pop." However, there was a crucial requirement for success. Botox would not have received FDA approval if patients occasionally died during the treatment—even if the injections could be shown to be effective. The dewrinkling deformations caused by the Ricci flow had better not lead to pathological situations either.

And this is where the problems start. A manifold might develop corners, grow constrictions, or split into several pieces. It could even go anorexic while seeking its ideal waistline. Yes, there may be side effects to the Ricci flow, and quite serious ones at that. (You may want to remember that botulinum toxin is no picnic either.) In mathematics, such pathologies are known as singularities.

Singularities began to pose problems for Hamilton. One instance that could be solved relatively easily was that the manifolds do not preserve their volume when the Ricci flow operates. The bulging parts lose, the dented parts gain volume. Some manifolds could be made to expand forever by the Ricci flow. Others disappear to a point so quickly that it is hard to see whether their curvature has become round like that of a sphere. This problem was overcome by continuously "renormalizing" the manifold, which means blowing it up or draining it just sufficiently to make its new volume match the original volume.

But other singularities could not be dealt with as easily. They occur when curvature in different directions develops at different speeds. By way of example, take a cylinder. It is round in one direction, straight in the other. Under the Ricci flow, scale decreases in the curved direction, making the cylinder curl up. On the other hand, scale increases where the cylinder is straight-edged, thus expanding it in that direction. The result is a tube that gets thinner and thinner and thinner. Eventually it will also disappear into thin air, but since the exact shape it had just before going "pop" was not round, it wasn't a sphere. On the other hand, shapes could look like tubes but might, in fact, be stretched-out spheres. However, if they go "pop" before their spherical shape can be verified, one cannot with confidence maintain that they are equivalent to a sphere. As in a murder trial based not on an admission of guilt but on a chain of evidence, the singularity's shapes must be observed uninterruptedly until the last moment of its existence. If anything, the requirement is even stricter in mathematics than in a court of justice. Any slight doubt, not just reasonable doubt, makes a guilty verdict impossible.

Consider a dumbbell consisting of two spheres connected by a handle—this is just a sphere whose midline has been squeezed. Since the handle's cross section is thinner than the spheres' diameter, and therefore has greater curvature, it will shrink faster than the rest of the dumbbell. After the Ricci flow has worked its magic for a while, the handle will have been squeezed infinitely thin and the dumbbell manifold will go singular: The spheres separate. This pathological situation is called the neckpinch singularity.

There is also the degenerate version of the neckpinch. It occurs when the two spheres of a dumbbell manifold are not equally large. If the sizes of the handle and the spheres are just of a certain proportion, the spheres do not separate under the influence of the Ricci flow. Rather, both the handle and the smaller sphere are simultaneously reduced in size, until only the larger sphere remains, albeit with a small protrusion. Looked at under a magnifying glass, this protrusion has a distinct similarity to a nipple.

Then there is the "cigar singularity." It would prove especially galling. Picture the end of a cigar, before the tip has been snipped off. As we know, the surface of a cigar is a two-dimensional object. Now "multiply" it by a one-dimensional line to get a three-dimensional object suspended in four-dimensional space. What happens when the Ricci flow operates on the surface of a cigar, multiplied by a line? Since the cigar end is curved, curvature increases, the object becomes tighter and smaller and finally retracts completely. In the other direction, however, the line is flat and does not budge. So the object caves in along the two dimensions, but stays put in the other. It does not go "pop" like the soap bubble vanishing into thin air, but "plouf" like a balloon slit open along its side. What stays behind is a flat and empty canopy. So the Ricci flow makes the object collapse: It isn't quite gone, but it isn't quite there either.

Hamilton spent about ten years trying to sort out the singularities. Then he found a solution. Remembering the profession of his father, the surgeon, he opted for radical treatment: Cut out the offending piece and get on with life. In other words, let the Ricci flow perform its task until shortly before the singularity is about to occur. Stop. Cut out the unwanted piece. Glue swaths onto both ends of the remaining manifold to cap the lesions. Restart the flow. Should the manifold relapse and develop another singularity, redo the procedure. Perform this as often as required, if need be infinitely often. And do keep track of the excised pieces. As pathologists do with tumors, subject them to further analysis.

In 1995 Hamilton published a 140-page monograph entitled "Formations of singularities in the Ricci flow." In it he described the surgery approach to the problem of singularities. Flow, stop, cut, glue, flow, stop, cut, glue. Keep track of everything, examine all the pieces. Check if anything went "pop." Start over.

But then he got stuck again. The cigar singularity just would not succumb to surgery.

* * *

I met Hamilton in Zurich in June 2006. He had been invited to give the Pauli Lectures at the Swiss Federal Institute of Technology, commonly known by its German acronym ETH (Eidgenössische Technische Hochschule), and the science editor of my newspaper had asked me to cover the event. Hamilton's talks in Zurich were to be a dry run for the plenary lecture that he would give at the International Congress of Mathematicians in Madrid two months later. The evening of the first lecture, Italy played against Ghana at the Mondial, the World Soccer Championships in Germany. The evening of the second lecture, the Swiss played against the French in their opening game. Switzerland being a soccer-crazed nation, millions were glued to their television screens. In spite of that, the Auditorium Maximum, the largest lecture hall at the ETH, was crowded both times. Not a single seat was left; latecomers had to sit on the stairs. Appearing in shirtsleeves and using only handwritten slides, Hamilton gave inspiring performances. The only off-notes were his references to his computer as, believe it or not, Muffin. Well, we did say he is a cool guy, did we not?

At a dinner party in his honor at the Belvoir Park Restaurant near the shore of the lake of Zurich, Hamilton was friendly. But he would absolutely not talk about Poincaré's Conjecture to a journalist. "There's too much at stake," he explained. "I'm still checking the details, together with colleagues. Topologists who can't verify the proof by themselves will be relying on us and that's why I won't say anything until I'm absolutely sure." All he did was refer to his lectures. In them Hamilton described the Ricci flow and the problems with the singularities. He recounted that one day he was told by a colleague about papers that a mysterious Russian mathematician had posted on the Internet. They purported to be a proof of Poincaré's Conjecture. But there had been too many hoaxes and Hamilton was weary. Eventually he did take a look. To his surprise, he realized that this was no hoax. "This guy may have something," he told his colleague.

But let's not jump the gun, Muffin.

Chapter 12

The Cigar Surgeon

Hamilton was stuck. He had made good progress during the past two decades, but after much heroic effort he could advance no further. The cigar singularity was an obstacle that refused to budge. Then, as if from nowhere, a Russian mathematician by the name of Grigori Perelman appeared on the scene.

Perelman is an intensely private person. As was recounted in chapter 1, he prefers not to socialize, has rejected the prestigious Fields Medal, and avoids the press. He lives with his mother in a high-rise building in a drab neighborhood of St. Petersburg. In December 2005 he resigned his post at the Steklov Institute of Mathematics of the Russian Academy of Sciences. The mathematician Mikhail Gromov, from the Institut des Hautes Études Scientiques near Paris, who has worked closely with Perelman over the years, professes sympathy for his behavior. "He is very sensitive to ethical issues, both in science and in society in general," he explains in an e-mail message. "He is critical, in my view often justly, of the decline of ethical standards in the mathematicians' community. But since he is reluctant to get involved in a controversy, he limits his non-professional contacts with people." Another mathematician gushes, "Perelman is a fascinating personality. I admire his integrity and independence." Who is this mysterious man who—willy-nilly—is about to become one of the icons of twenty-first-century mathematics?

Already as a child "Grisha," as his family and friends called him, seemed destined for greatness in science. He was the older of two children of Jewish parents from St. Petersburg (Leningrad at the time). The father was an electrical engineer, the mother a math teacher. When Grisha was still very young, his father often challenged him with logic and math puzzles, and he became adept at solving them. He attended the city's Public School Number 239, an institute that had been founded ten years earlier for exceptionally talented children. Just one month after his sixteenth birthday he called international attention to himself for the first time. At the Mathematical Olympiad in Budapest he answered all six questions perfectly and received a gold medal with the maximal score of 42 points.

The same year, he applied for university. Even though a strict numerus clausus was enforced against Jewish students at that time, he easily gained admission to study mathematics at the Saint Petersburg State University. After obtaining his doctorate Perelman took a low-ranking and low-paying research position at the department of geometry and topology at the Steklov Institute. He then got into an argument with his superior, the geometrician Yuri Burago, and changed to the department of partial differential equations. Burago does not want to speak about the reason for the altercation. "Our differences are due to Perelman's difficult character," he said in a telephone interview. "As is generally known, this is often the case with brilliant personalities." But the spat at the institute did have at least one beneficial consequence: The experience Perelman gained in both geometry and differential equations would stand him in good stead later.

Grisha's younger sister, Elena, attended the same school in St. Petersburg as her brother and also became a mathematician. She eventually followed her father, who had separated from his wife and moved to Israel. There she obtained her doctorate at the Weizmann Institute of Science. Following her marriage, she moved to Sweden, where she works as a biostatistician at the Karolinska Institute in Stockholm.

After a couple of years at Steklov, Perelman went abroad, and for mathematicians, as for many scientists, this means America. He obtained a position at the Courant Institute in New York in the fall of 1992, then spent the spring of 1993 at the State University of New York at Stony Brook. Often he and a colleague, the Chinese professor Gang Tian, drove to Princeton to attend seminars at the Institute for Advanced Study. On one such occasion, Richard Hamilton had been invited to talk about Ricci flows. Even though the subject was not his immediate interest, Perelman had read Hamilton's papers, and when the talk was over, he went up to the speaker to ask for clarifications about a certain point. Hamilton was friendly and open, sharing results that, at the time, he had not yet published.

In the fall of 1993 Perelman received a two-year Miller Fellowship at the University of California at Berkeley. He had obtained sensational results in geometry, and many colleagues were already keenly aware of his talent. In particular he had solved a problem that had remained open for twenty years with a surprisingly short and elegant proof. Among insiders he was considered a coming superstar. Admittedly, one had to search hard to find his pearls because he often could not be bothered to write down his results in a publishable form. Either he felt that a proof was too unimportant and that there was no need to publish it, or he made do with simply informing his colleagues. As he was not seeking tenure or promotion, publishing yet another paper just for the glory was against his convictions. In the atmosphere of "publish or perish" that prevails at today's universities, Perelman was a refreshing exception.

At Berkeley, Perelman blended in perfectly with the crowd in some ways, but glaringly did not in others. He kept his hair long, which was not unusual, but he also kept his fingernails uncut, which was. This foible sprang from his conviction that nature had not intended hair or nails to be cut. His aversion to cars was especially incongruous in California. One day in a supermarket in Berkeley, he met an Israeli colleague, Zlil Sela, whom he knew from an earlier occasion. Sela, who had just arrived in California, was taken aside by Perelman in the aisle and had to listen for half an hour to an insistent lecture. The subject: Cars are unnecessary, should be avoided, and Sela should under no circumstances buy one.

The Israeli mathematician got to know Perelman as an accessible colleague who occupied himself intensively, but not exclusively, with mathematics. He was interested in many things, Sela recounted, was neither uncommunicative nor unsocial. He liked to be informed about political developments, especially in Israel, where his father lived. True to his aversion to cars, Perelman usually walked everywhere, carrying books in a backpack. He lived modestly, usually wore the same clothes, and saved money wherever he could, for an important reason. Part of his fellowship money he sent home, to assist his mother and sister in St. Petersburg. Whatever he could spare, he saved for later times.

His thriftiness could get him into perilous situations. One evening, he was walking with a German colleague on campus when a mugger appeared from the dark and demanded their cash at gunpoint. While his colleague quickly surrendered his wallet, the unwilling Perelman was laboriously pretending to look for his. Not wanting to hang around for too long, the mugger eventually took off, leaving Perelman's savings untouched.

At the International Congress of Mathematicians in Zurich, in 1994, Perelman gave a lecture that received much attention from the experts. Since he had little money and the Steklov Institute did not fund travel, the Swiss airline Swissair had offered him a free ticket. Two years later, the European Mathematical Society awarded him a prize...or nearly did. Every four years, ten promising young mathematicians are honored by the association. Anatoly Vershik, a respected older colleague at the Steklov Institute, had suggested Perelman to the prize committee, and they indeed found that Perelman was worthy of an award. But if Vershik had thought that he was doing young Perelman a favor, he was mistaken. Perelman refused the prize.

Vershik recounts how Perelman justified his refusal. The work for which he was to be awarded the prize was not yet finished, he maintained, and was therefore not suitable. Vershik tried to reassure him.

Thereupon, Perelman asked who the members of the jury were who had decided on the worthiness of the prize. When Vershik told him the names, Perelman maintained that these people did not understand his work and the prize was simply razzmatazz. All further attempts to get him to change his mind proved fruitless. On the European Mathematical Association's Web site Perelman is still listed as one of the winners, but the prize was never awarded.

While at Berkeley, Perelman started to become interested in Poincaré's Conjecture. Hamilton had come to the West Coast several times to give talks about the Ricci flow. In the lectures he stressed his belief that this differential equations approach could lead to a proof of Poincaré's Conjecture. All that really stood in the way was the cigar singularity. It represented a formidable obstacle, however. This time it was Perelman who, after one of the talks, disclosed some of his results to Hamilton. He had not yet bothered to publish them but believed that they could be helpful with the cigar singularity. Later Perelman would confide to one of the few journalists with whom he agreed to meet that he got the distinct feeling that Hamilton had not understood what he had tried to explain.

Low-dimensional topology was not Perelman's chosen field of expertise, but now that differential equations had entered the picture, he started taking an interest. Keenly aware of his capabilities, he was on the lookout for a problem that would measure up to his talent. A ninety-year-old puzzle that had stymied many first-class mathematicians throughout the twentieth century was just the sort of challenge he sought. Without letting anyone in on his secret, he began asking around about previous attempts to prove Poincaré's Conjecture. He told nobody why he had suddenly developed an interest in the subject, and since his primary interest was not in topology, nobody seemed to suspect his intentions.

With the end of his fellowship drawing near, Perelman bid farewell to his American friends and colleagues. He had received job offers from first-class universities such as Stanford and Princeton but had refused them all. Instead he returned to his homeland. Back at the Steklov Institute he practically disappeared from sight. In nearly total solitude, he began working earnestly on the Poincaré Conjecture. He supplemented his meager salary by the money he had saved in the United States. For six years he worked alone, telling nobody about his secret. From time to time, when he required information, he sent e-mails with specific queries to his colleagues. Everybody who had some slight contact with him during that time confirms that his questions were deep and that Perelman was brilliant. Once he spent the winter months in absolute solitude at a friend's dacha. The friend came only once in a while to bring food and heating fuel. Since Perelman did not have any obligations and did not have to teach—the Steklov Institute is a research institute—the privacy in the bitterly cold hut was just what he needed.

Nearly eight years after he had first embarked on the path, Perelman felt that the time had come. Without any expert having examined his work, let alone having read and studied it, Perelman was convinced that he had solved the problem. He sat down to write out a series of three papers that would solve the Poincaré Conjecture and even the more ambitious Geometrization Conjecture.

In the papers, Perelman followed Hamilton's program, but while Hamilton's modus operandi worked for only a restricted set of manifolds, Perelman extended the method so it would work for all possible manifolds. Basically, it consisted of letting the Ricci flow operate on a compact, simply connected manifold and observing it until singularities develop. The list of possible singularities had already been whittled down to a short list and classified by Hamilton. Apart from the problematic cigar, only two kinds of singularities existed: spheres and long, thin tubes. Spheres, being one of the basic building blocks postulated by Thurston, do not pose a problem and can simply be removed. Tubes, on the other hand, are more complicated. They are either connectors between two parts of the manifold, or they are appendages, attached to the manifold at one end and capped off at the other. In the Ricci flow one watches for the first sign of a developing tube. The pathological piece is extracted from the manifold by prophylactic surgery just before it goes completely singular. Surgery is performed by cutting these tubes off near their ends and then closing all orifices by attaching caps to them. After the spheres have been removed, the tube singularities amputated, and the openings capped off, the Ricci flow is restarted.

This sequence of operations is repeated as often as needed, possibly ad infinitum. At each stage, it is verified that the removed pieces correspond to one of Thurston's primafolds. When no more singularities develop, the remainder of the manifold is inspected to verify that it, too, is one of Thurston's eight building blocks. And then the work is done. Walking backward in time, one realizes that all the bits and pieces that have been removed from the manifold are of one of the types that Thurston had predicted. Hence, the original manifold is built out of Thurston's primafolds, and if it has a trivial fundamental group, it must be a sphere. Thus the Poincaré Conjecture holds.

A crucial part in this program is the Canonical Neighborhood Theorem. In it Perelman showed that every tightly curved region is similar to either a sphere or a tubelike singularity, as described above. The conclusion is that such regions can be nudged into the shape of a sphere or of a tube and then either morphed into a ball or amputated.

So spheres and tubes have been dealt with, but what about the cigars that caused so much pain? Hamilton had realized that his program could not proceed if cigar singularities developed, since these singularities resisted surgery. Therefore he fervently hoped that such singularities would never occur. But while faith and hope may play a role in medicine, in mathematics they are not enough. One needs proof. Perelman showed that the cigar, the most pathological of all pathologies, could be ruled out. It could never occur in a manifold's life. The story of how he did that is a historic gem.

First of all, Perelman needed some special tools. One cool trick that he employed to inspect the manifolds on their way to going "pop" had actually been around for decades and had already been used by Hamilton. Called parabolic rescaling, it resembles watching a movie of the manifold's development under a microscope, while focusing on a developing singularity. While the manifold shrinks, the speed of the movie is slowed down, and simultaneously the manifold is enlarged. Both magnifications take place continuously, but with a twist. When slowing down and zooming in, the dilation of the time scale is linear at each magnification—two, three, four times…. But distances are scaled down like the square roots of these magnifications.

Then Perelman devised a quantity that does not change when the manifold undergoes parabolic rescaling. He called this quantity entropy because its mathematical properties resemble entropy in statistical physics: In the same manner as the disorder of molecules increases as an object is heated, Perelman's entropy increases over time as the manifold is deformed by Ricci flow. Many mathematicians had been searching for a useful quantity that would be invariant under parabolic rescaling, but Perelman was the first to find it. He did this with another cool trick. Recall that the Ricci flow was inspired by the differential equation that describes the flow of heat. Perelman examined the Ricci flow in reverse—running backward in time—and studied how the "temperature" of the manifold would change. Imagine temperature going backward in time: A room starts out with a uniform temperature throughout and ends up being hotter near the radiator and colder near the window.

With his new concept of entropy, Perelman was ready to take on the cigar. Under the parabolic rescaling microscope, sphere singularities just look like immobile spheres—even when magnified—and tube singularities just look like motionless tubes. The cigar singularity, however, has a life of its own. It becomes rolled up more and more tightly. Eventually, unlike the other singularities, it collapses.

But then Perelman pulled the rabbit out of the hat. Using his concept of entropy and some intricate mathematics, Perelman showed that manifolds cannot become too tightly rolled up. Except for manifolds that "pop" and go extinct, there must always remain sufficient room for a little ball, like a pea caught between the sheets of the collapsing canopy. Hence, when viewed under parabolic rescaling, manifolds cannot collapse in the Ricci flow. The pea prevents a total cave-in. In the terminology of the previous chapter, manifolds can go "pop" but never "plouf."

The Local Non-Collapsing Theorem, as the theorem was henceforth called, is the central ingredient that was needed to deal with the cigar singularity. As we saw in the previous chapter, the rules of the Ricci flow predict the cigar's eventual collapse. On the other hand, Perelman showed that collapses are ruled out. Combined, these two facts imply that the appearance of cigars is a mathematical impossibility. To misuse Freud, "sometimes a cigar is just a cigar." If only Hamilton had discussed his fears with the famous psychoanalyst. Topological cigars are only figments of the mind and have no basis in the reality of Ricci flows. But it took the Russian mathematician to disperse these worries.

Now Perelman was ready to return to Hamilton's program of running the Ricci flow and removing singularities. The next challenge was to ask whether an infinite number of surgeries would have to be completed in a finite amount of time. Perelman showed that this was not so.

Under the Ricci flow, the parts of the manifolds with negative curvature are beaten out. (Those with positive curvature are beaten in, but we don't have to worry about these parts here.) Whenever a manifold is being beaten out, it bulges and its volume expands. On the other hand, parts of the manifold are amputated during surgeries. This makes the manifold's volume decrease. Perelman put these two facts together. First, he showed that during any finite period the gains in volume due to the beating-out are limited. Thus the manifold's volume cannot become infinitely large during that period. Second—remember the pea caught between the sheets?—each surgery removes a certain amount of volume. Hence only finitely many singularities can be removed before the manifold's volume would be reduced to zero. And since the volume cannot become negative, an infinite number of surgeries is not possible during a finite time.

So during any given time span, the number of surgeries must be limited. But could such time spans be added, end to end, so that the program would continue forever? This was a difficult point. To prove Thurston's Geometrization Conjecture, Perelman described a process that would allow the virtual surgeons to continue performing surgery infinitely often for endless time. This was not altogether satisfactory, and in a third paper Perelman proposed a shortcut that would lead to a partial result: manifolds with a trivial fundamental group would not need endless surgery. This shortcut is not general enough for *all* manifolds—so it does not prove Thurston's Geometrization Conjecture—but it does suffice to prove Poincaré's Conjecture. Perelman performed his feat using—believe it or not—soap bubbles.

As we know, soap bubbles suspended in, say, wire frames form surfaces that span minimal areas. Perelman used such minimal surfaces to measure a certain kind of object that may arise during surgery, called a neck. We now move away from subtle surgical procedures to coarser amputations. The neck is the place where Perelman cuts the manifold. Let us consider the manifold to be the mythological multiheaded Hydra and Perelman to be Heracles, who is charged with killing her. Whenever he chops off a head, the Hydra keeps growing new ones. Only she does not just sprout spherical heads, sometimes she sprouts whole new bodies with necks in between them. Had she just sprouted heads, Perelman would not have had a problem because spheres eventually go "pop." However, he really needed to prevent the Hydra from sprouting extra bodies.

Heracles would have prevented the Hydra from growing new bodies by scorching the stumps. Our real-life hero, in contrast, shows that it cannot even come to that. (That is why Heracles and Hydra belong to a mythological tale, whereas Perelman and the manifolds are real...well, at least Perelman is.) The areas of the chopped necks can get smaller and smaller, but the areas of the necks between two bodies must have a definite size. Eventually the Hydra does not have sufficient skin to grow new bodies. At worst, she is able to sprout just tinier and tinier heads. These are just little spheres that we do not care about because…yes, because they go "pop."

By showing that eventually the Hydra would stop sprouting bodies, Perelman was able to prove that even if Heracles had to keep chopping off heads ad infinitum, the topology of the Hydra would no longer change. Hence, it could be seen in finite time that the Hydra was in fact a sphere, thus proving Poincaré's Conjecture.

Let us take a step back and survey what Perelman did. First, he developed tools to spot approaching singularities. Second, he invented methods to choose the right moments for prophylactic surgery. Third, he showed that only a finite number of surgeries is required. Thus, he proved that a compact, simply connected manifold that is deformed through Ricci flow and has all singularities removed by surgery will, in the end, be just a collection of spheres. If we run time backward, and paste the spheres back together, the original manifold is seen to be, itself, a sphere. Voilà, Poincaré's Conjecture.

In the fall of 2002 and the spring of 2003, Perelman posted three papers, one by one, in an Internet archive that was especially created by researchers to allow speedy distribution of scientific results. Called the arXiv—the middle letter X denotes the Greek letter *chi,* which is how the word *archive* is spelled in the original—it is an online repository for papers that have not yet been published by, and often not even submitted to, a learned journal. Started in 1991, it contains papers in mathematics, physics, nonlinear sciences, computer science, and quantitative biology. In mathematics alone, about seven hundred papers are uploaded every month on average. Everybody can post to the arXiv. There is no refereeing, though lately an endorsement system has been instituted. After a deluge of rubbish papers, the arXiv's advisory board decided that at least one person who has previously posted to the arXiv must give his or her blessing to a paper before it can be put into the arXiv.

Of course, the paper could still be junk, and some postings are, but serious mathematicians would not chance the embarrassment of an incorrect submission. Before posting their papers mathematicians take great care to ensure that these do not contain errors. Usually, after proving a theorem, and before going public via the arXiv, mathematicians embark on a road show, giving talks to various audiences at math departments around the country. This allows them to test out their ideas in front of experts, get their comments, and correct possible errors before they hit the airwaves. Recall poor Colin Rourke's devastating experience in Berkeley? At least he was spared more public embarrassment. Others, less cautious, proceed directly to press conferences to announce their accomplishments and then have to cope with the fallout. Not so Perelman. Of course nothing would have been further from his mind than a press conference. On the other hand, he was so sure that his arguments were correct that he felt no need for a road show either. He also required no endorsement for the arXiv because he himself had posted there previously. The upshot was that nobody—not a single person—had seen his papers before he dropped them on the Internet.

And what if he had made an error, if his papers contained a teeny-weeny unproven assertion or an itsy-bitsy undeclared assumption? This would not have bothered him, he told the two journalists from *The New Yorker* who sought him out in St. Petersburg. At least the postings would have allowed others to find holes and plug them, he explained, thus improving human knowledge. How about that? If this was not selflessness of the highest order in the interest of science, nothing—short, of course, of Giordano Bruno's burning at the stake—ever was.

If you think Perelman's mode of disseminating his findings to the mathematics community a wee bit arrogant, consider this: Never—not once—is Poincaré's name mentioned in the three arXiv postings. Perelman does not even bother to claim that he had proved the Poincaré Conjecture. Let the reader figure this out by himself. Thurston's much more ambitious Geometrization Conjecture is alluded to in paper number 1, but only in passing. The first mention occurs, hidden away, on page 3: "Thus, the implementation of the Hamilton program would imply the geometrization conjecture for closed three-manifolds." And, on the next page: "Finally, in §13 we give a brief sketch of the proof of the geometrization conjecture." Just by looking at the abstracts of these papers nobody would have guessed what a momentous achievement Perelman presented on these pages.

Perelman's reluctance to make a big deal of his work does not stem from arrogance. As far as Poincaré's Conjecture is concerned, those to whom Perelman aimed his papers would know what they imply, and those who would not know should not be reading his papers in the first place. So why make a fuss about the conjecture?

The first arXiv submission, posted on Monday, November 11, 2002, at thirty-nine pages the longest of the three, is entitled "The entropy formula for the Ricci flow and its geometric applications." At the bottom of the first page where Perelman gives his affiliation as the St. Petersburg branch of Steklov Mathematical Institute, the footnote reads, "I was partially supported by personal savings accumulated during my visits to the Courant Institute in the Fall of 1992, to the SUNY at Stony Brook in the Spring of 1993, and to the UC at Berkeley as a Miller Fellow in 1993–95. I'd like to thank everyone who worked to make those opportunities available to me."

True to his somewhat exaggerated sense of honesty, Perelman was extremely scrupulous throughout the paper in giving credit—of the nonfinancial kind—to everybody who deserved it. In fact, to ten of the thirteen sections he added paragraphs, marked with asterisks, that give historical remarks as to who did what before him. Ten papers by Hamilton are listed in the references.

The paper's last, thirteenth, section is entitled "The global picture of the Ricci flow in dimension three." Just before the asterisked paragraph, Perelman concludes the section, and the paper, with the words "Thus the topology of the original manifold can be reconstructed as a connected sum of manifolds...." And thus ends the sketch of the proof of the Geometrization Conjecture, the most significant achievement of recent mathematical history. No swaggering, no gloating; just the fact.

Four months later, on March 10, 2003, Perelman posted the next paper, "Ricci flow with surgery on three-manifolds." It contains twenty-two pages and is more technical. A continuation of the first one, it corrects some inaccuracies and verifies most of the assertions that he made in §13, the "brief sketch of the proof of the geometrization conjecture." Again, five of Hamilton's papers are cited, including a ninety-seven-page article of 1997, in which Perelman found an unsupported statement on page 62. On July 17, 2003, Perelman submitted the final installment, "Finite extinction times for the solutions to the Ricci flow on certain three-manifolds." This is the paper we described earlier as a battle with a Hydra. It provides a shortcut toward proving the Poincaré Conjecture rather than all of Thurston's geometrization. On seven pages, Perelman shows that the Ricci flow makes certain manifolds go "pop" in finite time.

Apart from the stupendously new mathematics in the three papers, Perelman's command of the English language may be quite surprising. The papers read as if they had been edited by a professional, not written by a Russian who had spent the last eight years hardly speaking English to anybody. Consider the following ninety-two-word sentence: "Otherwise we remove the components of Ω which contain no points of Ω_p, and in every ϵ-horn of each of the remaining components we find a δ-neck of radius h; cut it along the middle two-sphere, remove the horn-shaped end, and glue in an almost standard cap in such a way that the curvature pinching is preserved and a metric ball of radius $(\delta')^{-1h}$ centered near the center of the cap is, after scaling with factor h^{-2}, δ'-close to the corresponding ball in the standard capped infinite cylinder, considered in section 2." Maybe not quite intelligible, and we'll take

Perelman's word for the accuracy of the math, but there is no argument with the syntax.

The papers were for written experts in geometric analysis, not for topologists. And even they would have to invest great effort to study them. In fact, it was to take eighteen man-years to figure out whether everything was correct. (As we shall see in chapter 13, three two-man teams would work for three years each.)

Perelman sent e-mails to some of his friends, bringing his postings to the attention of colleagues and acquaintances. He got straight to the point, wasting no time on pleasantries like "How have you been in the meantime?" or "How's the weather?" For example, Gang Tian, his former colleague at the Courant Institute, then at MIT, received the following message on Tuesday, November 12, one day after Perelman had submitted his first paper.

```
Date: Tue, 12 Nov 2002
From: Grigori Perelman
To: Gang Tian
Subject: new preprint

Dear Tian,
may I bring to your attention my paper in arXiv
math.DG 0211159.
Abstract:
We present a monotonic expression for the Ricci
flow, valid in all dimensions and without cur-
vature assumptions. It is interpreted as an
entropy for a certain canonical ensemble. Sev-
eral geometric applications are given. In par-
ticular, (1) Ricci flow, considered on the space
of riemannian metrics modulo diffeomorphism and
scaling, has no nontrivial periodic orbits (that
is, other than fixed points); (2) In a region,
where singularity is forming in finite time, the
```

injectivity radius is controlled by the curvature; (3) Ricci flow can not quickly turn an almost euclidean region into a very curved one,
no matter what happens far away. We also verify
several assertions related to Richard Hamilton's program for the proof of Thurston geometrization conjecture for closed three-manifolds,
and give a sketch of an eclectic proof of this
conjecture, making use of earlier results on
collapsing with local lower curvature bound.

Best regards,
Grisha Perelman

The intrigued Tian had not heard from Perelman for years. But he remembered him well and immediately downloaded the paper. "I immediately realized its importance. Grisha is a very sharp mathematician and does mathematics very carefully. I thought that there must be something very unusual in the paper since he was claiming a very, very big thereom." Just three days later, Tian sent off an answer to Perelman.

Date: Fri, 15 Nov 2002 20:59:23 -0500 (EST)
From: Gang Tian
To: Grigori Perelman
Subject: Re: new preprint

Dear Grisha,
I am reading your paper. It is very interesting.

Will you be interested in visiting MIT and give a
few lectures on this work?

Regards! Tian

In discussions among themselves, colleagues became more and more enthusiastic. Another few days after the first posting, a colleague sent an e-mail to Perelman asking whether it was true that his paper, and the ones that were to come, would constitute a proof of the Geometrization Conjecture. The answer was not verbose: "That's correct. Grisha"

Information about Perelman's feat went around the world like movie-star news on TV. In fact, it did make it to all the major outlets, and the headlines were nothing if not exuberant. RUSSIAN REPORTS HE HAS SOLVED A CELEBRATED MATH PROBLEM (*The New York Times*), MAJOR MATH PROBLEM IS BELIEVED SOLVED (*The Wall Street Journal*), MATHEMATICS WORLD ABUZZ OVER POSSIBLE POINCARÉ PROOF (*Science*), some of them raved. The electronic media were equally flattering: GREAT MATHS PUZZLE SOLVED (BBC), RUSSIAN MAY HAVE SOLVED GREAT MATH MYSTERY (CNN). MathWorld, the more circumspect online encyclopedia of mathematics, waited a little while, then reported POINCARÉ CONJECTURE SOLVED—THIS TIME FOR REAL. Within days of each arXiv posting, everybody was talking about Perelman's newest paper.

Even Hamilton took notice. At first, he had ignored the postings. As I recounted in previous chapters, incorrect proofs had abounded in the past, and mathematicians had become wary of such claims, even if they were formulated as modestly as in Perelman's posting. But after a little while, an intrigued Hamilton took a closer look. So did a lot of other people. With excitement growing, the mathematical community was eager to hear firsthand what the noise was all about. Perelman was invited to the USA to lecture on his purported proof. He willingly accepted.

The first set of lectures was scheduled for April 7, 9, and 11, 2003, at MIT. Many people had heard about the cryptic papers that had been posted by that mysterious Russian, and nobody wanted to miss out on history in the making. One of them was Rob Kusner, from the University of Massachusetts, who drove from Amherst to Cambridge with two of his students on this damp Monday afternoon. The three described the event in their math department's newsletter:

The MIT lecture theater is packed. Late arrivals sit on the floor or stand at the back. Hundreds of people—young students, old professors, and many New England–area mathematicians from all fields of mathematics—await the news. Grisha Perelman, a mathematician from the Steklov Mathematical Institute in St. Petersburg, Russia, wears a long beard and a dark gray suit. He paces before two large blackboards, waiting to deliver the lecture. It takes a few extra minutes for everyone to get settled. At last, Perelman is introduced by MIT's Victor Kac. Perelman tests the microphone: he says that he is not good at speaking linearly, so he intends to sacrifice clarity for liveliness. The audience is amused. Then he lifts a jumbo white chalk to the blackboard, and writes the definition of the Ricci Flow, which was introduced in 1982 by Richard Hamilton.

Speaking before such a distinguished audience was daunting and would have intimidated any lesser man. But Perelman was unruffled. He used no notes to deliver his talk. He wore pants, a pair of old sneakers, and a jacket over a zippered jersey. With his long hair and fingernails he did not make a dapper figure. "If he'd stood on Harvard Square with a paper cup asking for change, he would have looked like any other homeless guy hanging around," recounted a former student who had come to his alma mater especially to attend the lecture. But everything changed as soon as Perelman started talking.

Admittedly, it was no slick PowerPoint presentation. After putting the equation for the Ricci flow onto one of the two huge blackboards in the lecture hall, he hardly wrote anything else for the next forty-five minutes. At the end of the talk, the whole audience burst into applause. Kac thanked the speaker...for wasting so little chalk. The initial lecture was followed by two more talks over the next days. Perelman impressed everybody. He was eager to explain his new ideas, parried all challenges, and answered questions without hesitation. At a private dinner after one of the lectures, several people tried to convince Perelman to stay and work in the United States. He could not be tempted.

On April 16 it was Princeton's turn. The auditorium was filled with distinguished mathematicians of all ages. Andrew Wiles, the man who had proved Fermat's Last Theorem, sat in the third row; the Nobel laureate John Nash had taken a chair two rows behind. Usually only one or two dozen people show up at a seminar, but this time more than a hundred had gathered to listen to Perelman describe his groundbreaking work.

The following week, lectures were scheduled in the framework of the Simons Lecture Series at SUNY at Stony Brook. Michael Anderson, who had been involved in a different approach to solving Thurston's Geometrization Conjecture for years, had graciously invited Perelman to give these lectures. Entitled "Ricci flow and the geometrization of 3-manifolds," they encompassed three two-hour lectures, with a half-hour coffee break, during the first week, and another three informal sessions during the following week. The latter were reserved for questions and answers and for discussions. The event was preceded by preparatory lectures a few weeks earlier in which the interested public was given an introduction to Ricci flows.

When the day arrived—April 21, 2003—again more than a hundred mathematicians were present at the lecture hall at Stony Brook. During the lecture, and the two that followed, Perelman once more showed how sincerely interested he was in conveying what he had done, not just in basking in the growing glory. The second week of critical perusal and discussion went equally well. Christina Sormani, an associate professor at Lehman College and CUNY Graduate Center, who had done research on Ricci curvature, drove to Stony Brook daily to attend the lectures. She knew Perelman from when she had been a graduate student at the Courant Institute and remembered him as quite the rave in her subfield of mathematics when he was a visiting mathematician.

This time he was a rising, if unwilling, superstar. Sormani found him rather shy, wary of the media and even of topologists. He felt more comfortable with geometers and was attentive to questions of experts who already knew about Ricci curvature and Ricci flow. She was struck by how humble Perelman was. He never seemed eager for self-promotion and heaped credit on Hamilton. During one of the lectures, he called one of Hamilton's proofs "a really miraculous achievement."

Sormani took notes of the first week of lectures, and a colleague took notes of the second week. They were posted on the Internet and also given to Hamilton, who was scrutinizing Perelman's papers at the time. Perelman did not mind being grilled for hours every afternoon after his formal talks. "I asked him many questions, and he answered them in a detailed and precise way, often with accompanying diagrams," Sormani recalls. The only time he lost his cool was when a member of the audience asked a somewhat elementary question. He indignantly brushed it aside; if he had to explain Ricci flow from scratch, he could spend a week just describing Hamilton's results before even getting started on his own findings.

Still another week later, Perelman lectured at the Courant Institute of New York University. This time, reporters mingled with mathematicians in the audience, which did not please Perelman at all. He refused to answer their questions, and when they asked him to speculate about the implications of his work, he waved them aside. His temper also got a bit short. When a photographer's flash went off, he snapped, "Don't do that!" His dislike of publicity is the reason that hardly any photos exist of Perelman, except for two that were surreptitiously taken during his lectures at MIT and Princeton.

Perelman was celebrated at each of his appearances, and people had come from far and wide, but there was one notable absence. Richard Hamilton had not attended any of the lectures. He had not bothered to make his way to MIT, Princeton, or Stony Brook. He had not even deigned to drive from Columbia University at 116th Street in Manhattan to the Courant Institute in Greenwich Village. This disappointed Perelman, especially since he considered himself one of his disciples. Hamilton's apparent reluctance to learn about the new development in his chosen field of expertise was understandable, if not excusable. He had known for a long time that he was stuck and had openly solicited help. But after having spent the better part of two decades trying to solve the Poincaré and the Geometrization Conjectures, just to see a long-haired Russian wander in from obscurity to snatch away the ultimate prize, must have hurt, nonetheless.

John Morgan from Columbia University decided to give it another try. He scheduled a previously unannounced additional lecture for Saturday morning at his math department, where Hamilton also taught. This time, finally, Hamilton did show up, albeit late. He did not ask a single question, neither during the discussion time nor during the luncheon afterward. But a physicist who had been sitting next to him during the talk wrote later in his blog that Hamilton was clearly impressed.

The rest is history. During the following three years, nobody found even one serious error in Perelman's three postings to the arXiv; obscurities maybe, sketchy sections that needed more explanations certainly, but no errors. As reported in the opening chapter of this book, Perelman was awarded a Fields Medal at the International Congress of Mathematicians in August 2006. He refused it.

Where is Perelman today and what does he do? The Russian daily *Izvestia* described him as a scientific hermit. *The New Yorker*'s special envoys did manage to talk to him, and to them he seemed quite accessible. But he generally avoids even colleagues and leaves e-mails unanswered. With his resignation from the Steklov Institute, he severed his last official connection to academia and no longer owes anybody any explanation. People who know him say the best way to meet him is to go to the opera in the hope of running into him. Opera is one of the few interests he pursues. But the theater that is being made about him and his proof is repugnant to Perelman. He seems to be motivated only by the search for mathematical truths. Some rumors have it that he has retired from mathematics altogether. But is he maybe, just maybe, on to something else? Sormani ventures, "I suspect Perelman may be working much, much more than fifty hours a week, even if he pretends he is no longer doing mathematics at all."

Chapter 13

The Gang of Four, plus Two

Having finished his whirlwind tour of the USA and disappeared again into the Russian wilderness—actually to the city of St. Petersburg—Perelman left behind a plethora of new tools and methods that was to keep many mathematicians busy for years. No less than three two-man teams started working intensively on his legacy. They were John Morgan from Columbia University and Gang Tian from MIT, Bruce Kleiner and John Lott from the University of Michigan, and Huai-Dong Cao from Lehigh University in Pennsylvania and Xi-Ping Zhu from Sun Yat-Sen University in China. The Clay Mathematics Institute, a not-for-profit organization in Boston "dedicated to increasing and disseminating mathematical knowledge," had set aside a million-dollar grant to study Perelman's proof. Richard Hamilton also rolled up his sleeves and got down to work, together with Gerhard Huisken from the Max Planck Institute of Astrophysics near Berlin, in Germany, and Tom Ilmanen from the ETH in Zurich, Switzerland. For three years the teams read, studied, and dissected every page and every word that Perelman had posted to the Internet. They labored until the summer of 2006.

Perelman's papers were dense and terse. They contained only the essentials; much was left unsaid. In general, an advanced paper is no textbook; authors take for granted that the basic facts are known to the intended readers. Though advances in science always build upon the achievements of earlier scholars, one cannot start every paper with Adam and Eve. Mathematics is no exception. Papers that contain, say, differential equations do not mention Newton's and Leibniz's invention of the calculus.

But even taking into account mathematical customs, Perelman's Internet postings were extremely concise. His refusal to write up a detailed and more accessible paper has since been described by some colleagues as either egotistical or lazy. What would Perelman's reaction have been if one of the other great mathematicians of our day had sent him a half-written idea asking him to check it? one frustrated researcher asked rhetorically. Maybe then he would have understood the daunting difficulties that all those interested in his papers had to face.

In late August and early September of 2004, the Gang of Four (Morgan, Tian, Kleiner, and Lott) ran a two-week workshop at Princeton University, sponsored by the Clay Institute. The participants studied Perelman's papers in detail. He had obviously made extremely important advances toward answering the Poincaré Conjecture, but whether his proof was correct was still an open question. The Princeton meeting would prove crucial. The attendees at the workshop went through the papers with a fine-tooth comb. Even though the correctness of everything could not be established with finality, it became increasingly clear that Perelman had not made any major blunders. "This workshop played a significant role in convincing us that Perelman's arguments were complete and correct," Morgan and Tian would later write.

The following year, the Gang of Four felt that they had progressed far enough to present Perelman's proof to a broader audience. With the support of the Clay Mathematics Institute and the Mathematical Sciences Research Institute (MSRI, sometimes pronounced *misery*) they ran a summer school on Ricci flows in Berkeley in June and July 2005. The program was designed for graduate students and mathematicians within five years of their Ph.D. It was organized "around Ricci Flow and the Geometrization of 3-manifolds, particularly, the recent work of Grisha Perelman." The Gang of Four, augmented by Hamilton and others, served as lecturers.

Morgan had met Perelman during the lecture tour in 2003. He found the visitor from Russia socially ill at ease, but forthcoming and patient when it came to mathematics. His talent and incredible insights became apparent immediately. Tian had known Perelman in 1992 at the Courant Institute of Mathematical Sciences at NYU, when he was a young professor and Perelman a Russian postdoc. Tian had become aware of the first arXiv posting after he'd received the e-mail from Perelman in November 2002.

At first, Tian only wanted to see if anything in the papers could be of benefit for his research. So, trying to learn Perelman's new techniques, he organized a seminar at MIT in which Perelman's work was studied in detail in the spring of 2003. Then he taught a one-year course on Perelman's works at Princeton. Little by little, Morgan and Tian became convinced that Perelman had actually hit the jackpot. Furthermore, they both found the papers beguiling in their beauty. But somebody needed to check the nuts and bolts. It was already apparent that Perelman was not going to produce any more papers to explain his findings. "I didn't want the argument to just hang there," Morgan said in an interview a few yeas later. And Tian added, "Perelman's paper is very sketchy and hard to read for nonexperts. John and I thought that it would be very useful to the mathematical community if we wrote a book with a detailed proof. This made good sense to us since by September of 2004 we knew his papers very well."

As a service to the mathematical community, Morgan and Tian took the chore upon themselves. They limited their objective to a precise question: Did Perelman prove Poincaré's Conjecture, and if so, how? The proof of Thurston's Geometrization Conjecture would be left to Kleiner and Lott. But even with that limited objective, the effort grew to proportions they had not imagined at the outset. "Had we known how much work it would be, we would maybe have decided differently," Morgan would say after the work was done.

They toiled for close to three years. When they hit a snag, they often sat there wondering, "How did this guy do it?" Messages were rushed off to Perelman via his e-mail address at the Steklov Institute. The answers were always fast, precise, and to the point. Obviously, Perelman wanted his ideas to be understood. The more Morgan and Tian came to grips with the proof, the more they became impressed by its sheer elegance.

But it was rough going throughout. Tian and Morgan discussed among themselves, consulted with Hamilton and other colleagues, and sent more queries to St. Petersburg. Sometimes it took weeks until a subtle point was finally understood. The more they dealt with Perelman's work, the less its conciseness bothered them. "If I had only one paragraph to say something," Morgan commented, "and given what I now understand, I would have written it exactly as Perelman did." There were no excess words, but nothing essential was missing either.

A year after they started work on their project, the contacts between Columbia University, MIT, and Princeton (where Tian had moved in the meantime) on the one side, and the Steklov Institute on the other, started to dwindle, finally ceasing altogether. By now, Morgan and Tian had understood the argument, and there was no longer any need to communicate on urgent mathematical matters. Given Perelman's reluctance to exchange friendly chatter, the e-mail exchanges stopped. "Perelman isolated himself from the mathematical community," Morgan sighed. "It was his choosing, I wish it were otherwise."

In contrast to Perelman, who had written for the experts, Morgan and Tian aimed their work toward graduate students and general mathematicians interested in seeing how the Poincaré Conjecture had been conquered. "Because of the importance and visibility of the results discussed here," they wrote in the introduction, "and because of the number of incorrect claims of proofs of these results in the past, we felt that it behooved us to work out and present the arguments in great detail.

Our goal was to make the arguments clear and convincing and also to make them more easily accessible to a wider audience." So, while Perelman's papers ran to a total of sixty-eight pages, Morgan and Tian's exegesis was seven times as long.

The first five chapters, covering 125 pages, contain introductory material and explanations to make their piece self-contained. But even so, this was no "Poincaré for Dummies." In the remaining 348 pages Morgan and Tian delved into the mathematical depths. In July 2006, their 473-page paper, "Ricci flow and the Poincaré Conjecture," was posted to the arXiv. Sponsored by the Clay Mathematics Institute, it is to be published as a book. Morgan and Tian refuse to take any credit whatsoever for having done anything of significance beyond expounding on Perelman's ideas. They did no more than unpack and reorder his findings, they emphasize.

During the three years that Morgan and Tian pored over Perelman's papers, Kleiner and Lott labored on their own version of a commentary. It was to be entitled "Notes on Perelman's Papers" and, with eventually 192 pages, would be substantially shorter, but no less rigorous, than Morgan and Tian's book. Their focus was the proof of Thurston's Geometrization Conjecture.

To keep interested parties abreast of any new developments, Kleiner and Lott maintained a Web site at the University of Michigan. It was a repository for scholarly material devoted to the Ricci flow and Perelman's work. Various new versions of their own paper were posted from time to time, enabling friends and colleagues to comment.

Perelman's remarkable proofs were concise but sketchy, Kleiner and Lott remarked at the outset of the "Notes." Their task, as they saw it, was to fill in the gaps. "The purpose of these notes is to provide the details that are missing," they wrote, and, like Morgan and Tian, produced a self-contained piece of scientific workmanship. "Besides providing details for Perelman's proofs, we have included some expository material in the form of overviews and appendices." But don't get your hopes up. Even with the inclusion of expository material, reading Kleiner and Lott's "Notes" is no walk in the park. Quite aware of the material's difficulty, the authors took pains to warn potential readers. Their "Notes" are definitely rated "PG: professional guidance recommended." They were intended for mathematicians "with a solid background in geometric analysis."

While Morgan and Tian's work was laid out as a book that would contain all the details and be accessible to graduate students and general mathematicians, Kleiner and Lott did not mean their "Notes" to be self-contained. They were supposed to be a companion to Perelman's papers and to be read in conjunction with them. Closely following the organization of Perelman's work, they provide a punch-by-punch or, rather, a line-by-line, section-by-section exposition and explanation. Only rarely is some of the material shuffled around a bit. They encountered numerous difficulties and even some errors. "Perelman's papers contain some incorrect statements and incomplete arguments, which we have attempted to point out to the reader," they write. But none of these gaps were even remotely serious enough to invalidate any part of the proof. Perelman himself corrected some errors that had slipped into his first paper in the second posting. All others could be fixed without the need for any additional tools or methods. "We did not find any serious problems, meaning problems that cannot be corrected using the methods introduced by Perelman." Hence, according to Kleiner and Lott's well-considered opinion, Perelman had provided all the necessary ingredients for the proof of Thurston's Geometrization Conjecture. They posted the "Notes" to the arXiv in May 2006, two months before Morgan and Tian posted their version.

All the while, unbeknownst to many mathematicians, two more scholars were trying to make heads and tails of Perelman's postings, the two Chinese mathematicians Huai-Dong Cao and Xi-Ping Zhu.

The story of their verification of the proof of Poincaré's Conjecture is not quite as straightforward as the endeavors of the Gang of Four. In fact, it became a story of intrigue and high drama. Accusations and allegations were thrown about, reproaches and recriminations abounded. Even a lawyer had a walk-on part. The affair became ugly, but ever so enticing for blogs and even mainstream media around the world.

Born in 1959, Cao had graduated from Tsinghua University in Beijing in 1981 and obtained his doctorate at Princeton University in 1986. He was the second of forty or so graduate students of Shing-Tung Yau, the mathematician who had been so instrumental in inspiring Hamilton. Yau also aroused Cao's interest in the Ricci flow. After completing a postdoc at Columbia, receiving a Sloan Fellowship, and teaching at Texas A&M, Cao was appointed professor at Lehigh University in 2003. In 2004 he became a Guggenheim Foundation Fellow. Throughout his career he produced significant results on Ricci flow.

Zhu was much less well-known outside China. He graduated from Sun Yat-Sen University in 1982 and received his M.A. from the same university two years later. He obtained his doctorate in 1989 from the Wuhan Research Institute of the Chinese Academy of Science. His work on partial differential equations was well recognized in China, and in 2004 he was a recipient of the Morningside Medal, a high distinction awarded every three years to Chinese mathematicians under the age of forty-five. He is now a professor of mathematics at his alma mater, Sun Yat-Sen University.

When Perelman's postings suddenly appeared on the Web, Yau was taken aback. He had always believed that Hamilton's program would eventually lead to success. At the same time, he was keenly aware of the remaining problems. So when Perelman's papers appeared on the Internet, they came as a complete surprise. But the postings to the arXiv were not clear, and Yau expected that more detailed explanations would be forthcoming. When they were not, he turned to Zhu and to his former student Cao and suggested they go through Perelman's proof. The two talented mathematicians worked for nearly three years, producing a one-and-a-half-inch thick manuscript. In their endeavor, they mastered many hurdles. Strangely, they never contacted Perelman for help. Rather they devised ways around some of the difficulties that the Gang of Four had also encountered.

Finally they were satisfied that they had a full proof of the Poincaré

Conjecture. To test-run it and straighten out any remaining obscurities, Yau organized a seminar at Harvard. Every week from September 2005 until March 2006, Zhu met for a grueling three-hour session with senior mathematics faculty. During tough discussions and incisive questionings, he managed to dispel any doubts, thus convincing the participants that Cao and he had, in fact, written down a valid proof. In December 2005, while the seminar was still in full swing, the paper was submitted to the *Asian Journal of Mathematics*. It was duly accepted by the journal's editorial board.

That is the story in a nutshell. But under the apparently bland surface, tempers began to rise.

The article "A complete proof of the Poincaré and geometrization conjectures—Application of the Hamilton-Perelman theory of the Ricci flow" was to fill the whole June issue of the journal, covering 326 pages. When the table of contents of the *Asian Journal of Mathematics*' forthcoming issue, containing only the single item, was made public, Chinese media were overjoyed.

An announcement by Sun Yat-Sen University under the heading SYSU PROFESSOR ZHU XIPING SOLVED ONE-HUNDRED-YEAR-OLD MATH MYSTERY reads as follows (verbatim, including the somewhat idiosyncratic grammar):

> Ponicare [*sic*] Conjecture, one of the most famous open problems in mathematics and has been around for about one hundred years, has been completely solved by scientists recently. Renowned mathematician Qiu Chentong [Yau, Shing-Tung], a Harvard professor and Fields prize winner, announced in CAS Chenxing Math Center on June 3 that Professor Zhu Xiping with Sun Yat-sen University and Professor Cao Huaidong with Lehigh University in Pennsylvania had produced a complete proof to the conjecture on basis of research achievements by preceding American and Russian scientists. "This can be compared to the construction of a building. The forerunners have laid the foundation,

while the last phase work, 'topping out,' is completed by the Chinese," said Qiu Chentong. "It's an extraordinary accomplishment, much more important than Goldbach Conjecture."

Xinhua, the Chinese news agency, was quick to pick up the story: "Beijing, June 4—A leading Chinese mathematician, Yang Le, said here Sunday that the successful unraveling of one of the world's toughest puzzles is an outstanding job.... Two Chinese mathematicians, Zhu Xiping and Cao Huaidong, have put the final pieces together in the solution to the puzzle that has perplexed scientists around the globe for more than a century.... Perelman raised guidelines for proving the conjecture but not specifically pointed out how to unravel the puzzle." The *People's Daily* reported TOP MATHEMATICIAN RECOGNIZES CHINESE WORK ON SOLVING POINCARÉ CONJECTURE. News traveled as far as India. The India eNews Web site reported SOLVING TOUGHEST PUZZLE/OUTSTANDING JOB: CHINESE MATHEMATICIAN. And Chinese embassies around the world exulted CHINESE MATHEMATICIANS SOLVE GLOBAL PUZZLE.

All this media attention shocked the mathematical world. It seemed that Cao and Zhu were claiming the credit for themselves. Nobody had expected another proof, and a claim that Chinese scholars had solved the Poincaré Conjecture seemed preposterous. The Gang of Four had known nothing. Or did they just feign ignorance? It is hard to believe that a semester-long seminar at Harvard would go unnoticed. Nevertheless, neither Kleiner and Lott's Web site nor their paper in May made any reference to the impending publication of Cao and Zhu's article. Morgan and Tian do mention the new paper in their posting in July, but their surprise is palpable: "After we had submitted a preliminary version of this manuscript for refereeing, H.-D. Cao and X.-P. Zhu published an article on the Poincaré Conjecture and Thurston's Geometrization Conjecture."

So was there really an attempt being made to usurp the glory? Or was Cao and Zhu's genuine effort to clarify Perelman's proof willfully misinterpreted and ignored? Gang Tian definitely knew about Cao's work because he had been coauthoring with Cao recently. Cao had also been presenting his work at Columbia. Furthermore, Kleiner-Lott were well aware of the Cao-Zhu paper but refused to list it on their Web page. The mathematical community was abuzz. To gain an understanding of the developing controversy, we must take a close look at Cao and Zhu's paper.

The two authors do seem to give credit where credit is due, mentioning Hamilton and Perelman's work right at the outset. "This proof should be considered as the crowning achievement of the Hamilton-Perelman theory of Ricci flow," they write in the abstract. Now it may be due to a misunderstanding by Asian scholars not quite at ease with the English language, but "this proof" conveys the subtle impression that this here is, finally, a proof, while everything that came before was not quite a proof. And "crowning achievement" refers to the ones who provided "this proof," not to those who provided the theory of Ricci flow.

We move on. The paper's first paragraph ends with the laudatory remark "The major contributors are unquestionably Hamilton and Perelman," which is, however, put into perspective by the immediately preceding sentence. It reads, "We shall give the first written account of a complete proof of the Poincaré Conjecture and the geometrization conjecture of Thurston."

In the following paragraph, criticism is veiled but the intention shimmers through. "We would like to point out that our proof of the singularity structure theorem is different of that of Perelman.... These differences are due to the difficulties in understanding Perelman's arguments at these points." Now, this can mean one of two things: either Cao and Zhu had difficulties in understanding the arguments because they are not smart enough, or Perelman's arguments are incomprehensible because they are incorrect. Since it is unlikely that anybody would admit to his less than perfect intelligence, the unwritten implication of this utterance is clear. Cao and Zhu are, supposedly, the ones who provided the correct proof.

Elsewhere Cao and Zhu lament that Perelman promises a proof that "is still not available in the literature," thus forcing them to do the work themselves. "We will provide a proof...by only using [a different] result.

In particular, this gives another proof of the Poincaré conjecture." Still elsewhere they claim that Perelman did not explicitly prove a certain statement but that they managed nevertheless. "We are still able to obtain a weaker version of [the statement] that is sufficient to deduce the geometrization result." Are they implying that they provided their own proof, after having shown Perelman's to be incomplete?

By now the intention should be clear. But for those who have still not got the message, the authors become even more explicit toward the end of the introduction. "As we pointed out before, we have to substitute several key arguments of Perelman by new approaches based on our study, because we were unable to comprehend these original arguments of Perelman which are essential to the completion of the geometrization program." And just to make sure that everybody's contribution gets the recognition it deserves, but not more, they point out to the reader that the notes written by Kleiner and Lott "cover part of the materials that are needed for the geometrization program."

The introductory chapter ends with profuse thanks to Yau, who introduced the authors to the "wonderland" of Ricci flow, suggested the paper, and provided vision, many suggestions, and consistent encouragement. "Without him, it would have been impossible for us to finish this paper." The authors are also "enormously indebted to Professor Richard Hamilton for creating the Ricci flow and developing the entire program." Perelman is not on the list.

Their credit to Perelman is more hidden. Some of it comes in the middle of the introduction: "Perelman brought in fresh new ideas to overcome the main obstacles that remained in the program of Hamilton." Their work "grew out of the effort" of the mathematics community to understand "whether the proof of Poincaré Conjecture and the geometrization program, based on the combination of Hamilton's fundamental ideas and Perelman's new ideas, holds together." In the more mathematical part of the introduction they refer to Perelman's work as "spectacular," and throughout the article they attribute specific observations, statements, and proofs to Perelman. Nevertheless, many mathematicians who glanced only at the introduction and had heard the reports by the Chinese press became too angry even to read the paper.

The article was submitted on December 12, 2005, and accepted on April 16, 2006. This is a blinking warning light for any historian of mathematics. A time span of four months would barely suffice to referee a regular midsize mathematical paper. To seriously appraise a 326-page paper in such a short time is quite impossible. The second light blinks upon inspection of the editorial board of the *Asian Journal of Mathematics*. Shing-Tung Yau, Cao's Ph.D. adviser and Zhu's mentor, is listed as its editor in chief. A third light goes on when one realizes that other manuscripts that had been accepted for publication by the journal in February had to wait for the September issue while Cao and Zhu's was pushed ahead of the line and published in June...barely six weeks after its acceptance. Clearly, it was timed to appear before Perelman's Fields Medal would be announced at the ICM in Madrid.

Even though the personal relations between the paper's authors and Yau were well-known, the decision to publish Cao and Zhu's article was made without consulting the twenty-six members of the journal's editorial board. They were not shown the paper, not even an abstract, and no reports by independent referees had been commissioned. In fact the decision to accept the paper was steamrolled through by the editor in chief, with the members of the editorial board notified of the pending publication just a few days before the journal issue appeared. What makes the matter worse in the eyes of many in the mathematics community is that none of the editors seem to have protested. Even in the affair's aftermath—when all the facts were out in the open—not one of them deemed it appropriate to resign his position.

Upset by the breakdown of commonly accepted norms, Joan Birman from Columbia University vented her sadness and anger to the *Notices of the AMS*. "As mathematicians, we have an extraordinary tolerance of eccentricity," she wrote in a letter that appeared in January 2007. "But there is another, and a darker, side to the same phenomenon, i.e., a tolerance for bad behavior, especially when the individuals whose actions might be questioned are highly talented. To put it plainly, we do not police ourselves very well." In the affair of the Cao-Zhu paper "the normal peer review process, essential to the integrity of the profession, was tossed out the window." Her sad conclusion was that "as a result the entire profession has received a very public and very bad black mark."

Two weeks after publication of the Cao-Zhu article, from June 19 to 24, the international physics conference "Strings 2006" took place in Beijing. It was organized by Yau, whose mathematical work had been instrumental in advancing string theory. The scientific meeting opened to great fanfare in the Great Hall of the People. The legendary icon of theoretical physics Stephen Hawking gave the opening speech in front of six thousand people. On the evening of day two of the conference a lecture was scheduled that was advertised alternatively as "Professor Yau's special talk" or as "Professor Yau presents his new research results." When the time approached, the conference hall was filled to capacity; apart from the conference participants, Chinese guests had been invited. Using two overhead projectors, one with the English version of the slides, the other with the Chinese version, Yau gave a first-class PowerPoint presentation: "Ladies and gentlemen, today I am going to tell you the story of how a chapter of mathematics has been closed and a new chapter is beginning. Let me begin with some elementary observations."

He then proceeded with a popular introduction to the Poincaré Conjecture. Everything went well until the two main actors, Hamilton and Perelman, should properly have addressed the audience. From that moment on the presentation degenerated. Of course, Perelman was nowhere near the Great Hall of the People. It is doubtful that he, back in St. Petersburg, even knew about the gathering in China. So Yau had someone read quotes from Perelman's papers. They were designed to show how much he, Perelman, appreciated Hamilton's work and that he had actually only carried out his, Hamilton's, program.

Then it was time for Hamilton's remarks. Only he was not in Beijing either. He had been there before the conference, but had left by the time

Yau gave his talk. So Hamilton addressed the audience through a video recording that had been filmed earlier. In the rather amateurish recording, the lighting was bad, there were sudden cuts, some parts were played twice. In his remarks, Hamilton stressed the spectacular mathematical results that Yau had proved, the brilliant students that he had taught, and the important role that Chinese mathematicians play in differential geometry. His talk ended, "All Chinese can be proud of the achievements of their mathematicians in differential geometry and their great contributions to the completion of the proof of the Poincaré Conjecture."

Three quarters through his lecture, Yau came to what he called Perelman's breakthrough. Characterizing his contributions as important and crucial, Yau clearly considers the proof of the Poincaré and geometrization conjectures to be due to Hamilton's program, a task that was brilliantly completed by Perelman. He ended his presentation, however, with what seems to have been his main concern throughout: "In Perelman's work, many key ideas of the proofs are sketched or outlined, but complete details of the proofs are often missing. The recent paper of Cao-Zhu, which was submitted to *The Asian Journal of Mathematics* in 2005, gives the first complete and detailed account of the proof of the Poincaré Conjecture and the Geometrization Conjecture. They substituted several arguments of Perelman by new approaches based on their studies."

Yau could not resist a swipe at two of the Gang of Four: "Kleiner and Lott [in 2004] posted on their Web page notes on some part of the work of Perelman. However, it was far from complete. After the work of Cao-Zhu was announced by the journal in April 2006, Kleiner and Lott put up another, more complete account of their notes. The approach is different from Cao-Zhu. It will take some time to understand their notes, which seem to be sketchy at several important points."

To Yau's dismay, the bizarre video screening was caricatured by some of the conference participants after the conference dinner the next evening. The persiflage took the form of a movie based on the previous evening's video, complete with bad lighting, strange camera angles, and effusive praise for some imagined scientist. The inside joke went a bit off target since those who had not been at Yau's talk had no idea what it was all about, while those who had been there were not quite sure whether they were allowed to laugh. Afterward, Eva Silverstein from the Stanford Linear Accelerator Center, who had been among the instigators of the movie, had mixed feelings. "Our video was simply a light joke. It was fun to do because the technology required was simply a laptop we had along, so we could plan and execute this joke...in time to show it during the after-dinner speeches." But she regretted the unintended problems it may have caused. "Almost by definition, a joke such as this is not a serious issue," she wrote Yau half a year later. Yau himself was not amused. Asked what he thought about the after-dinner performance, he responded that this was a kind of humor he did not understand.

On June 19, a science reporter at *The New York Times*, Dennis Overbye, dispatched a report from Beijing entitled HAWKING TAKES BEIJING; NOW, WILL SCIENCE FOLLOW? Two days later, the *International Herald Tribune* carried the piece under the title CHINA MOVES TOWARDS CENTER OF THE COSMOS. The article reports on the leading scientific role that China means to take in the future. To enable its scientific establishment to achieve this goal, the Chinese government intends to raise spending on research to about 2.5 percent of the gross domestic product.

Herein may lie the explanation for Yau's efforts to gain recognition for his protégés. In the mid-1990s, Yau had met the general secretary of the Communist Party and the president of the Chinese Republic, Jiang Zemin, several times. The subject of their discussions was the resurrection of the country's scientific institutions that had been destroyed during the Cultural Revolution. The scientific revival of his country is important to Yau, and he is undoubtedly passionate about this noble cause. The achievements of Chinese mathematicians would serve as a springboard for realizing this ambition. What role does the dyed-in-the-wool American Richard Hamilton play in the greater scheme of Chinese things? Yau is an old friend who had helped, supported, stimulated, and encouraged him for decades. Hamilton simply responded in kind. If that meant going a little overboard with praise for Chinese mathematicians, then so be it.

The verdict of the Gang of Four is unanimous. When asked by a reporter at the ICM in Madrid about the Cao-Zhu paper, Morgan answered, "It seems to me an honest account with the appropriate credit all the way through, and all the controversy stems not from this paper but from the noise around it." However, he maintains that, apart from clarifications, Cao and Zhu did not add anything of substance to the proof. They may have simplified some parts of the proof, but they did not fill any conceptual gaps. Kleiner and Lott also ascribe the proof exclusively to Hamilton and Perelman. Cao and Zhu had substituted arguments of their own for the parts that they did not understand; in a similar manner Kleiner and Lott had filled in some of the gaps that Perelman had left open. Clearly, Yau thought it a very impressive feat that his younger Chinese colleagues had absorbed it all and presented a self-contained account. But neither their paper, nor Kleiner and Lott's, nor Morgan and Tian's deserves special status. In terms of priority all three of them are also-rans.

Hamilton himself is a bit more lenient. In his ill-fated video talk in Beijing he asserted that Cao and Zhu had introduced ideas of their own that made the proof easier to understand. That may be the most decent statement, fair to all sides. But if Cao and Zhu simplified the proof, or even if they had filled some gaps of exposition, does that mean they should be credited, at least partly, with the proof? Not at all. No scientific progress occurs in a vacuum, every advance builds on previous work. Isaac Newton was not just being poetic when he said, "If I have seen further it is by standing on the shoulders of Giants."

But neither giants nor also-rans get credited every time their work is utilized. Only if Cao and Zhu had shown that one of Perelman's statements was wrong, say by providing a counterexample, and then demonstrated that their own formulation does not suffer from the same defect would their paper have to be credited as an original contribution.

Otherwise it simply belongs to the collection of simplifications and expositions, as does the work of the Gang of Four.

While all this was going on, *The New Yorker* published a piece on Perelman and the Poincaré Conjecture. Apart from a highly interesting interview with Perelman, the authors, Sylvia Nasar and David Gruber, also investigated Yau's involvement and his possible motivations. They painted a highly unflattering picture of an aging, conceited, and power-hungry professor, battling for his standing and reputation. The fourteen-page article was accompanied by a caricature of Yau in which he tries to pull the Fields Medal off Perelman's neck. The article appeared on August 22, 2006, the opening day of the ICM.

Yau was, once again, not amused. Especially, the caricature aroused his ire. It was actually extremely unfair, since Yau had done his best to convince everyone that Perelman would be a worthy recipient.

And what does Cao think about the brouhaha that has been aroused around his and Zhu's paper? "Unfortunately, some of the media's attention seems to be much more interested in something else than discussing mathematics. I simply focus on the mathematical part," he maintained at the International Congress of Mathematicians in Madrid. "In my joint paper with Zhu we have given a detailed account. Hamilton and Perelman have done the most important fundamental works. They are the giants and our heroes! ... We just follow the footsteps of Hamilton and Perelman and explain the details. I hope that everybody who read our paper would agree that we have given a rather fair account."

Nice as these words sound, an eerie feeling may befall the reader that the lady protesteth just a bit too much. Given the wording of some parts of the article and the behavior of some of those involved, one may be just a bit hesitant in assigning all the blame to the media.

The unamused Yau had his lawyers draw up a thirteen-page rebuttal and direct threats of a lawsuit toward *The New Yorker*. He also became active on the PR front. On September 16 a Web site was created (doctoryau. com) that went live two days later with the headline HARVARD MATH PROFESSOR ALLEGES DEFAMATION BY NEW YORKER ARTICLE; DEMANDS CORRECTION. On

September 20 a Webcast followed in which the attorney read large parts of the rebuttal for all the world to hear. It culminated in the cry "Dr. Yau wants his reputation back." A week later, Hamilton weighed in at doctoryau.com with a two-page character-witness piece: "I am very disturbed by the unfair manner in which Shing-Tung Yau has been portrayed in the *New Yorker* article. It is unfortunate that his character has been so badly misrepresented." More postings followed, and by the end of 2006 the Web site displayed no less than fifteen letters from colleagues asserting what a great guy Dr. Yau is.

Then an embarrassing event occurred. Working his way through Cao and Zhu's paper, Sujit Nair, a postdoc at the University of Southern California, had the feeling that parts of it seemed strangely familiar. Upon closer inspection, he realized that his feeling was well founded: Some pieces had been lifted verbatim from the initial version of Kleiner and Lott's paper. Well, well, well. The paper that had been lambasted by Yau in Beijing as being "far from complete" had served as a cribbing platform for his protégés. All of a sudden, the proud statement "We shall give the first written account of a complete proof of the Poincaré Conjecture" sounded hollow.

Soon thereafter a contrite erratum appeared on the *Asian Journal*'s website "We would like to thank Bruce Kleiner and John Lott for bringing to our attention that the argument...essentially appeared in the initial version of their notes." What happened was that while working on Perelman's arXiv postings, Cao and Zhu had been studying Kleiner and Lott's earlier paper, taking copious notes for themselves. Two years later, when they eventually wrote up their paper for the *Asian Journal*, they based themselves on their notes, having forgotten how these had originated. "We apologize for failing to attribute this argument to Kleiner and Lott in our paper due to an oversight," the erratum read. "We would like to acknowledge here that our treatment, with some modification, follows the June 2003 version of the notes of Kleiner and Lott."

On December 13, 2006, a paper was posted to the arXiv by Cao and Zhu. It carried the title "Hamilton-Perelman's Proof of the Poincaré

Conjecture and the Geometrization Conjecture." A footnote explained, "This is a revised version of the article by the same authors that originally appeared in the *Asian Journal of Mathematics*." In the new paper's abstract, the bombastic "We shall give the first written account" is replaced with the much more humble "We provide an essentially self-contained and detailed account" and the phrase "a complete proof of the Poincaré Conjecture" is expanded into "a complete proof of the Poincaré Conjecture due to Hamilton and Perelman."

The opening paragraph says it all: "In this revision, we have also tried to amend other possible oversights by updating the references and attributions. In the meantime, we have changed the title and modified the abstract in order to better reflect our view that the full credit of proving the Poincaré Conjecture goes to Hamilton and Perelman. We regret all the oversights [that] occurred in the prior version and hope that this revision will right these inattentions." Thus a whole new purpose was invented for the arXiv. Up to then, papers had been posted to this Internet repository so that colleagues could point out errors and omissions *before* the articles were submitted and published in conventional journals. This time a posting was made to change the content and set the record straight *after* the paper was published. If only Cao and Zhu had taken this route from the beginning, a lot of ill will and bad blood could have been avoided.

In January 2007 the American Mathematical Society's 113th annual meeting and the Mathematical Association of America's 90th annual meeting were held jointly at two posh hotels in New Orleans. One of the highlights planned was a special event on the Poincaré Conjecture and the Geometrization Theorem. But the event was not to be. The session had to be canceled when it became apparent that John Lott would refuse to sit at the speakers' table together with Xi-Ping Zhu. A lapidary announcement on the AMS Web site stated, "We regret that the special event on the Poincaré Conjecture and Geometrization Theorem has been canceled. It became apparent that the continuing controversy was undermining this special event."

Yau weighed in once more with a reader's letter published in the April 2007 issue of the *Notices of the American Mathematical Society*. Cao and Zhu's paper had been "accepted according to the standard editorial procedure of the journal, by which acceptance was automatic unless an objection was voiced within a few days of the chief editor's recommendation," he wrote, thus unintentionally revealing more about the autocratic goings-on at the *Asian Journal* than he probably had wanted to. But as the following sentence shows, the subtlety seems lost on him anyway. "This procedure of the *Asian Journal of Mathematics* of requiring consent from the whole editorial board is more stringent than [the procedure at] several leading mathematical journals where the chief editor would consult only a few members closest to the subject of the paper." Analyzing these statements mathematically, Yau apparently believes that "objection by nobody" is tantamount to "consent from everybody" which, in turn, is preferable to "consulting a few." Let the reader be the judge. A few lines later he maintained that he sees nothing wrong with expediting the reviewing process "for important solicited papers," which is the first time we are informed that Cao and Zhu's paper had actually been solicited by their mentor.

Seeing all this hullabaloo, can anybody really fault Perelman with wanting to keep his distance?

Priority disputes are as old as science itself, and mathematics is no exception. Three hundred years ago, Isaac Newton and Gottfried Leibniz argued fiercely about who had invented the calculus. And even today historians of science disagree whether Albert Einstein or David Hilbert first wrote corrected, submitted, and published the crucial paper on the theory of general relativity. But something good does arise from priority disputes. If scientists did not worry about who will be first, they would go about their research at a leisurely pace. And this would certainly not be beneficial to the advancement of science.

The ICM in Madrid in August 2006 provided the appropriate endpoint to the century-old saga. Perelman had, of course, stayed away. Shortly after King Juan Carlos left the conference center, John Lott gave the laudation speech for Perelman's Fields Medal. In the afternoon, Richard Hamilton was the first plenary lecturer. He had also finished checking the proof, together with Gerhard Huisken and Tom Ilmanen. There were the handwritten slides again, and the reference to Muffin, his computer. He explained the theory behind it all, his travails over the years, and Perelman's achievement. Hamilton did not forget to mention Zhu and Cao. But his greatest praise was reserved for the absent Perelman. "Thanks to Grisha, we'll never have to worry about 'noncollapsing' again," he exclaimed at one point, and ended his talk with the words "I'm enormously grateful to Grisha for finishing this all. I'm really happy."

A few days later, it was John Morgan's turn. His talk was billed as a special lecture. "Grigori Perelman has solved the Poincaré Conjecture," he announced at the start, and applause erupted. "Many before him foundered on the rocks. This is a signal achievement not only for Perelman, but also for mathematics." He recounted that in the 1970s mathematicians had still been split more or less evenly between those who believed the Poincaré Conjecture to be true and those who thought it was not. Ten years later, without any real advances and with no more evidence, the community's opinion had moved ten to one for Poincaré's Conjecture. "Now, finally, it is no longer just a matter of opinion," he concluded.

Most mathematicians believe not much will change now that the Poincaré Conjecture has been proven. "There is just a great sense of relief," Morgan said. Not even the name will change, he thinks. Even though Perelman has turned the conjecture into a theorem, in the popular mind it will always remain the "Poincaré Conjecture." Many topologists feel a wistful sadness, however, a sort of postpartum depression. The great adventure that had seen so many ups and downs since its inception in 1904, had kept hundreds of mathematicians busy for a century, and had made and nearly ruined many a career has now, finally, come to an end.

Chapter 14

The Prize

D id I say that the International Congress of Mathematicians in Madrid was the endpoint of the saga? Well, it wasn't quite. There is still the minor matter of a million dollars that have been promised to the person or persons who proved the Poincaré Conjecture.

Given the singular dedication that most mathematicians bring to their chosen field of study, this monetary reward is probably the least important aspect of the story. The truth is that mathematicians don't prove theorems for the money, and some don't do it for the glory. For many, the successful solution of a problem is reward enough. However, "a million dollars" does capture the imagination. And that was the intent of the donors, after all.

Prizes were an important part of the scientific environment of the eighteenth and nineteenth centuries. They enticed budding as well as established scholars and shaped the direction of research. Poincaré's Oscar, described in chapter 4, is a case in point. The academies of sciences in Paris, Berlin, and St. Petersburg were the centers of learning at the time. In 1719, Paris began the tradition of setting questions and awarding prizes every two years. Berlin started its prize competitions in 1745. Some of the most famous names in mathematics participated in the competitions. From 1725 onward, Daniel Bernoulli won ten prizes from the Académie des Sciences in Paris, one jointly with his father, Johann, and one jointly with his brother Johann II. Leonhard Euler won no less than twelve prizes from various academies. Joseph-Louis Lagrange won the French prizes in 1764, 1766, 1772 (jointly with Euler), 1774, and 1778. Simon l'Huilier, whom we met in chapter 5, won the Paris prize in 1786 for his entry on "a clear and precise theory of what is called infinity in mathematics."

At that time, prizes were prospective, in the sense that they were announced with a prescribed title and a deadline. This allowed the administrators of the prize, renowned scientists themselves, to determine which fundamental problem should be on the research agenda for at least the next couple of years. Today's Nobel Prizes and the Abel Prize for mathematics are retrospective. They are rewards for work done, often during the laureate's entire lifetime. The Fields Medals are somewhere in the middle. On the one hand, they are awarded for work already accomplished. On the other hand, the recipient must be young, the expectation being that he or she will perform outstanding work in the future. Since I did say "he or she" just now, let me point out that there were two outstanding female prizewinners in the nineteenth century. Sophie Germain won in 1815 in spite of a scandal when one of the judges, Siméon-Denis Poisson, who had seen an earlier version of her work, entered a paper himself. And Sofia Kovalevskaya was awarded the Prix Borodin, administered by the French Academy of Sciences, in 1888.

Not all competitions were successful. In 1815 the French academy proposed Fermat's Last Theorem as a Borodin Prize question. There were no takers. After an extension of two years—and still no entries—the question was withdrawn. The experiences of the Berlin academy during the nineteenth century were even more embarrassing. In 1836 a question about numerical methods for solving polynomial equations was asked. When no answers were received after four years, the question was replaced by another. Again there were no entries. At that stage, the academy decided to take some time out. Then, in 1852, another attempt was made. Again no entries. In 1858 a different question was posed and one submission arrived. Alas, it was quite unsatisfactory. Six years later, the exasperated jury tailored a question to the work of one of the judge's students, and finally, in 1866, they could award a prize. At this stage the academy again decided it was time to lie low for a while. When the prize was offered in 1894 with a question on differential equations, guess what? No entries. Four years later the question was asked again in different form, but there were still no answers. Disgruntled, the Berlin academy awarded the prize to a mathematician for "contributions to mathematics." The original question was rephrased but could still not be answered. As a last resort, Paul Koebe was awarded the prize in 1910 for his uniformization theorem (see chapter 11). And a new question was posed. There were no answers yet when World War I started, and from then on the members of the academy had other matters to worry about.

On August 8, 1900, at the International Congress of Mathematicians in Paris, David Hilbert—along with Henri Poincaré, the world's leading mathematicians of their time—delivered one of the most famous speeches on a mathematical topic that was ever given. He announced a series of twenty-three unsolved problems whose solutions he believed to be of supreme importance. No prizes and no medals were offered, but the challenge sufficed. His list defined the path that mathematics took for at least the first half of the twentieth century and kept hundreds of mathematicians busy for decades. A few of Hilbert's problems, for example the Riemann Hypothesis, are still not settled today. The Poincaré Conjecture, proposed only in 1904, was of course not on his list.

The idea of a modern version of prizes for mathematics came to Arthur Jaffe, the Landon T. Clay Professor of Mathematics and Theoretical Science at Harvard University. Jaffe had received a well-rounded scientific education. He obtained a bachelor's degree in chemistry in 1959 from Princeton, a BA in mathematics—as a Marshall Scholar—from Cambridge University in England two years later, and completed a doctorate in physics, at Princeton again, in 1966. He has written more than 160 scientific articles, mainly in quantum field theory; authored or coauthored four books; edited seven more; served as chair of the mathematics department at Harvard, as president of the International Association of Mathematical Physics, as president of the American Mathematical Society; received the Dannie Heinemann Prize in mathematical physics, a medal from the Collège de France; has been named a member of the National Academy of Sciences. You get the picture. He is also fluent in German and often visits colleagues in Germany and Switzerland. He enjoys good company with colleagues, and one of his hobbies is photography. To my surprise, I found a photo of myself on his Web site one day.

The idea of establishing prizes for mathematical achievements came to Jaffe in two stages. The honor of being the incumbent of an endowed chair comes with the custom of having lunch with the donor every couple of weeks. This donor, Landon T. Clay, was born in 1926, attended Harvard College in the late 1940s, and received his AB degree in 1950. He went on to become a multimillionaire, building his wealth as the chairman of Eaton Vance, a Boston-based mutual-fund giant. Over the years, shrewd investments allowed him to cash in hundreds of millions of dollars. Among his other ventures is the ADE Corporation, a firm that makes equipment for computer-chip manufacturers, of which he serves as chairman. Although nothing is made public about his remuneration, the annual take-home pay of his underlings may give an indication. In 2005, ADE's chief exectuve officer pulled down north of $800,000. Obviously, Landon T. Clay has no financial worries in the world and can devote a substantial part of his vast wealth to philanthropy.

And this he does. He is chairman of the Caribbean Conservation Corporation, of which his mother had been a founding member. This institute operates a turtle-nesting station in Costa Rica. It tracks turtles, named Señorita Chiriqi or Suzy Snowflake, by satellite and instituted an adopt-a-turtle scheme to lure more donors. Clay donated a telescope to a Harvard program in Chile, had science centers built in two high schools, and serves on various boards, most notably the Museum of Fine Arts in Boston. To its permanent collection of Mayan jade, pottery, and burial urns, he donated some valuable pre-Columbian objects. A controversy arose around this donation when an archaeology professor from Boston University claimed that the artifacts may have been looted from graves and exported illegally. Nobody accuses Clay himself of having done anything untoward, but on that occasion the savvy investor may have been taken in by grave robbers.

Of course he did not forget his alma mater, and there are not one but two Landon T. Clay professorships at Harvard, the one for mathematics and theoretical science that Jaffe holds, and another for scientific archaeology. Clay also created a substantial fund to assist the dean of arts and sciences in recruiting new faculty. In general, he likes the sciences and believes that they are a driving force for the betterment of mankind. But his overriding affection belongs to mathematics. Although he had obtained his degree in English, he had taken some calculus courses as an undergraduate. He had found them intellectually stimulating, and a lifelong fascination for the discipline ensued.

Clay knew from his experience in various Harvard committees, however, that the math department was underappreciated by the university administration. This is by no means a phenomenon particular to Harvard, as Jim Carlson, the University of Utah mathematician who was named director of the Clay Mathematics Institute after Jaffe stepped down in 2003, points out. Mathematics just does not receive the same support as do, for example, biology or physics. Mathematicians at universities all over the United States have higher teaching loads, for example, than their colleagues in the chemistry, physics, or biology departments. Carlson maintains that this is a direct reflection of the differences in funding.

So, in addition to all his other benevolent activities, Clay donated generously to the mathematics department. When the cold war ended and Soviet mathematicians started traveling, he funneled $4 million into a fund that allowed the department to invite visitors and enable joint research projects.

During one of Clay and Jaffe's luncheons, Clay brought up the idea of creating a foundation devoted to software. Jaffe, who had toyed with the idea of a mathematical institute, pointed out that such a foundation would have to compete with large software companies, which made little sense. It would be much more effective, financially and in impact, to create a foundation devoted to mathematics. A few months later, Clay made up his mind. A mathematics institute it would be, independent from Harvard and with Jaffe as the director of the scientific board. On September 25, 1998, the Clay Mathematics Institute (CMI) became a nonprofit corporation in the state of Delaware. Its Statement of Purpose read, "The primary objectives and purposes of the Clay Mathematics Institute, Inc., are to increase and disseminate mathematical knowledge, to educate mathematicians and other scientists about new discoveries in the field of mathematics, to encourage gifted students to pursue mathematical careers, and to recognize extraordinary achievements and advances in mathematical research. The Clay Mathematics Institute will further the beauty, power, and universality of mathematical thinking."

To this Jaffe added in an interview with the *Newsletter of the American Mathematical Society*, "We also hope to inspire people outside of mathematics to appreciate the importance of the field." The board of directors of CMI was initially composed of Landon T. Clay, his wife, Lavinia, and Jaffe. Soon thereafter three additional members were added: Finn M. W. Caspersen, William R. Hearst III, and David B. Stone. Like Landon and Lavinia Clay, they are enormously wealthy and prominent icons in the American business community.

Jaffe co-opted three outstanding mathematicians to serve on CMI's scientific advisory board. They were the Frenchman Alain Connes, who won the Fields Medal in 1982 together with William Thurston and Shing-Tung Yau; the American Edward Witten (Fields Medal, 1990), renowned for his work in mathematical physics; and the Englishman Andrew Wiles, who proved Fermat's Theorem and missed getting the Fields Medal only because he was already forty-one years old when the International Congress of Mathematicians met in 1998.

Half a year later, with the lawyerly work completed and the paperwork out of the way, the founding was celebrated with a grand mathematics conference at MIT. Since then, CMI has fulfilled its mission by supporting scholars, funding research projects, organizing summer programs, and publishing research results.

But the program for which CMI is best known among the public is the creation of the Millennium Prizes. That was the second stage of Jaffe's endeavor. In his original memo to Clay, in which he outlined his vision of a mathematics institute, he had already tossed up the idea: "In association with the millennium, I recommend a monetary prize for the solution of a small number of outstanding, long-range mathematical problems." It was an appropriate point in time. Hilbert had formulated his problems for the twentieth century exactly a hundred years earlier.

At first, Jaffe thought of collecting fifty problems in an open contest, with the fifty winners being paid $1,000 each. After that, up to a dozen or so problems would be chosen from among this list for the grand prizes. But this procedure would become too cumbersome, and the scientific advisory board decided to make do with at most a dozen problems from the outset. Jaffe, Connes, Witten, and Wiles would select suitable candidate problems, possibly after consulting confidentially with trusted colleagues. The problems would have to be difficult, important, and time-tested. The search was to be kept confidential to avoid mathematical politics and to keep important mathematicians who might attempt to influence the board members at bay.

The four members of the scientific advisory board started out by making four initial lists. The Riemann conjecture and the Poincaré Conjecture were on everybody's list and were thus immediately accepted as suitable Millennium Problems. In the following months more problems were added, some discarded after discussion, and more added again. By the end of 1999, the board members agreed on a list of seven questions and decided to stop. Jaffe admits that a different board might have come up with a different list. But he describes the selection as an "honest attempt to convey some of the excitement about mathematics." The winners were the Birch-Swinnerton-Dyer conjecture, the Hodge conjecture, the Navier-Stokes equations, P versus NP, Yang-Mills theory and the mass gap hypothesis, the Riemann hypothesis, and, of course, the Poincaré Conjecture.

The scientific advisory board asked specialists to write up an account for each problem. The recorder for the Poincaré Conjecture was John Milnor from SUNY at Stony Brook, who had won a Fields Medal in 1962 for his work on seven-dimensional spheres. By the time the official book on the Millennium Prize Problems came out, in 2006, Milnor's chapter had to be revised to reflect the new developments. It is the only chapter that needed revision. The status of the six other problems has remained unchanged.

In 1998, Steve Smale had proposed a list of eighteen problems for the twenty-first century. Some of them were the still unsolved, original Hilbert problems; some others, such as Poincaré's Conjecture, are identical to the Millennium Problems. However, since his list was not backed by a well-endowed institute, it did not generate as much attention as the list of Millennium Problems.

The next question the scientific advisory board had to decide was how to frame the problems. For example, should the Poincaré Conjecture be listed as the Millennium Problem or the more general Geometrization Conjecture? They decided that the problems would be phrased in their simplest form, which in this case meant the less generalized version of this topological conundrum. Little did the scientific advisory board know that Perelman was already more than halfway along to solving both of them in one fell swoop.

Next, the question of a reward had to be settled. All that had been decided was that it needed to be substantial. In England, the publishing house Faber & Faber had just announced a prize of $1 million for the proof of, or a counterexample to, a famous problem, in order to publicize the book *Uncle Petros & Goldbach's Conjecture* by Apostolos Doxiadis. So CMI also had to make a splash. Little did it matter that Faber & Faber had limited their offer to an unrealistic two-year period—the Goldbach conjecture has been around since 1742—which allowed them to take out insurance from Lloyd's against the improbable event that someone would really prove it or provide a counterexample within that time. The Clay prizes were open-ended, with real cash to be paid sometime in the future.

Jaffe's initial idea was to arrange for a prize fund that increased, albeit slowly, each year. So a problem solved after sixty or more years would yield a substantial reward. Then, taking a cue from Faber & Faber, he suggested $1 million each for the seven prizes. On the one hand, $7 million in prize money was bound to attract attention. On the other hand, the problems were sufficiently difficult to ensure that the cash would not have to be shelled out all at once or, indeed, for a long time. Mr. and Mrs. Clay accepted, and the Million Dollar Millennium Prizes were born.

Unfortunately no adjustment for inflation is provided. And that could prove to be a problem. After all, who knows when the Riemann conjecture or the Hodge hypothesis will be proven? It could take dozens, if not hundreds, of years. In 1906 the German mathematician Paul Wolfskehl offered a reward of 100,000 German marks to anybody who would prove or disprove Fermat's last theorem. At the time this was an enormous amount of money, corresponding to about $2 million in today's money. But by the time Andrew Wiles had proved the theorem, inflation had eaten up most of its value. (Remember that there were two world wars and at least one period of hyperinflation in between.) When the Wolfskehl Prize was awarded in 1997, it was worth only 75,000 deutschemarks, which was equivalent to about $50,000. Not that that really mattered to Wiles, but it could become a minor disappointment to the eventual winner of a Millennium Prize. Carlson expressed his confidence in the dollar's strength. "I believe the prize will keep its value," he said in early 2007. However, he did concede that the corresponding bylaws could be changed by the Clay Institute's board of directors sometime in the future.

With the amount of the prize decided, the time had come to determine the rules of the game. How would the scientific board decide whether a proof was valid? After all, the famous Four-Color Problem was believed proven in 1879 and again in 1880, before both proofs were shown to be incorrect eleven years later. And Andrew Wiles thought that his proof of Fermat's Theorem was complete until gaps appeared during refereeing. It may take several years to spot a subtle error, and only the test of time can tell whether a theorem's proof is correct. The assumption in the "test of time" is that if a proof contains a hole, eventually someone, somewhere, will spot it. The related assumption is that if during a reasonable time nobody, nowhere, finds any holes, there aren't any.

Though this does not completely, absolutely, 100 percent guarantee a watertight, rigorously correct proof—a tiny crack, overlooked by the myriads of professionals, may still appear sometime in the future—this is as close to absolute truth as one can reasonably get. As a consequence, the scientific advisory board decided on a three-stage test. First, the proposed proof to a Millennium Prize problem would have to be published in a well-respected mathematics journal. Thus, the obviously erroneous papers will already be weeded out by the journal's refereeing. The second stage consists of a two-year waiting period. It was designed to leave the proof open for inspection by the general community of mathematicians. If by the end of the period no serious objections have been brought up, the proof will be considered prima facie valid.

After the two-year waiting period, the third stage begins. The scientific advisory board appoints a panel of experts, specialists on the problem, who inspect all relevant material and then write a report. Based on this report, the scientific advisory board makes a recommendation to the directors of the Clay Mathematics Institute, who will then make the final decision on awarding the prize and the million dollars.

The ceremony to announce the Millennium Prizes was held in the summer of 2000. Of course, it had to take place in Paris, to tie in with the famous lecture, exactly a hundred years earlier, by David Hilbert in the same city. The vast amphitheater at the Collège de France was filled to capacity. To accommodate the many who could not find seats in the main hall, the proceedings were transmitted by closed-circuit television to a nearby room. Lectures presenting the seven Millennium Problems were given by Sir Michael Atiyah, former president of the Royal Society, and John Tate from the University of Texas. The only mishap was that the French minister of research did not show up on time and the schedule had to be rearranged on the fly.

The prizes hit the front page of Le Monde, France's leading newspaper, an Associated Press report was carried by several hundred U.S. newspapers, and *Nature* published an editorial. Once the prizes had been announced, pandemonium broke loose. Everybody and his uncle wanted to get in on the action; would-be prizewinners urgently needed to know where to submit solutions. The Web server that hosted CMI's Web site crashed, and when the site was mirrored to the much more powerful server of the American Mathematical Association, traffic threatened to crash that too. After a while, traffic to the CMI Web site leveled out, and catastrophes were averted.

During the first year following the announcement, mathematics journals were deluged with crackpot papers. Amateurs, drooling over the promised prize money, started sending in their manuscripts to learned journals. Not realizing the vast difficulties, they claimed to have solved a Millennium Problem, and sometimes all. An editor of the *Journal of Number Theory* complained to a reporter of *The Boston Globe* that he was getting more crank stuff than legitimate papers. "They are really coming out of the woodwork," he lamented. Eventually this flurry too abated.

Not everybody agrees that monetary prizes are an appropriate vehicle to further general interest in mathematics. In the January 2007 issue of the *Notices of the American Mathematical Society*, the distinguished Russian mathematician Anatoly Vershik from the Steklov Institute in St. Petersburg argued that the unfortunate clamor and fuss surrounding the proof of the Poincaré Conjecture was in large part due to the prominence that the Clay prize had bestowed on it. This just went to show, according to Vershik, "that this method of promoting mathematics is warped and unacceptable, it does not popularize mathematics as a science, to the contrary, it only bewilders the public....Does mathematics need such an indecent interest?"

He continues, "To transform serious research problems into something like a million-dollar lottery is a totalistic [*sic*] means to indulge the bad taste of the mob." As a case in point, he remarks that the Clay Institute had played no role whatsoever in furthering or hastening the solution of any

of the seven Millennium Problems. Those who had wrestled with Poincaré's Conjecture had been interested in the problem long before the million-dollar prize was announced.

Vershik has a valid point. But his disdain for earthly pleasures also be-
trays an elitist view of which mathematicians have often been accused. After all, it was not the aim of the prize to fund research. "You understand
nothing of the American way of life," Jaffe told Vershik on one occasion.
Prizes with large amounts of money attached were not meant to channel
more talent toward a specific problem, but to popularize mathematics
among the general public so that parents would "not discourage their children from choosing the profession." Vershik remained unconvinced: "Pop-
ularization of math for the general public is indeed necessary, but not of
the kind that is characteristic of the worst manifestations of present-day
mass culture." The utterance, which echoes Perelman's pure sentiments,
calls to mind the commendable, if naïve, idealism of the Soviet system in
which both Vershik and Perelman were brought up.

The opinion of detractors notwithstanding, the Clay prizes are here to
stay. So let us now examine the small print. The key phrase in the official
announcement of the Millennium Prizes reads, "Before consideration, a
proposed solution must be published in a refereed mathematics publication of worldwide repute, and it must also have general acceptance in the
mathematics community two years after." Note that the rules do not say
"the author must publish" his or her proof. The implication is that someone other than the inventor of the proof may publish it, and it would still
be the true creator who would get the credit. In fact, Jaffe had in mind a
famous Hilbert problem solved by Andrew Gleason but never published
by him. Rather Deane Montgomery and Leo Zippin wrote a book based
on Gleason's lectures. The credit for the solution of the problem is nevertheless always given to Gleason.

What if the purported proof is not published in a journal, such as
Perelman's? Or what if journals won't be around by the time that proofs to
the Millennium Problems are found? After all, the Poincaré Conjecture

was a hundred years old before Grigori Perelman came along, the Four-Color Problem two hundred when it was proven, Fermat's conjecture three hundred, and Kepler's conjecture four hundred. In December 1999, I asked Robert MacPherson from the Institute for Advanced Study what question he would ask his colleagues if he went to sleep and woke up at the start of the fourth millennium. His answer: "Has the Riemann conjecture finally been proven?" So it could easily take until the twenty-second or twenty-third or even the thirty-first century to prove or disprove some of the Millennium Problems. By then, scientific periodicals as we know them today may no longer exist. Who knows what means of disseminating research results will prevail a couple of centuries hence?

When the scientific advisory board began to foresee such possible developments, the members added a proviso to the official rules. At the end of the sentence "Before consideration, a proposed solution must be published in a refereed mathematics publication of worldwide repute," the phrase "or in such other form as the scientific advisory board shall determine qualifies" was added. This leaves the back door open for means of publication other than standard journals.

Finally, there is the question of how to allocate the prize once a proof has been recognized as prizeworthy. The rules specify that "the scientific advisory board will pay special attention to the question of whether a prize solution depends crucially on insights published prior to the solution under consideration." This means that the board may recognize prior work and recommend splitting the prize among multiple solvers. Or it may just mention it in the laudation. Or it may ignore it altogether.

All these questions apply to the case of Perelman and the Poincaré Conjecture. At this point it seems unlikely Perelman will ever submit his proof to a traditional academic journal. Does the arXiv qualify as "such other form" of publication? Maybe. Does Cao and Zhu's publication in the *Asian Journal of Mathematics* satisfy the requirement? Again, maybe. If the members of CMI's scientific advisory board decide that Perelman's arXiv submissions of 2002 and 2003 constitute a valid form of publication, then the two-year waiting period has long since elapsed without any errors or gaps having been found. In this case, the Clay Mathematics Institute could start its own procedure immediately.

Even if the arXiv postings do not represent a valid form of submission by the rules of the Millennium Prize, the publication of Cao and Zhu's paper, and the eventual appearance of Morgan and Tian's book, could fulfill the requirements. In this case the advisory board would have to wait until at least June 2008 before starting its procedure. On the other hand, if the members of CMI's scientific advisory board decide that the author himself must publish his work, then it is conceivable that the prize for the proof of the Poincaré Conjecture will never be awarded. Carlson believes that the scientific advisory board will eventually settle on the date of the announcement at the ICM in Madrid as the beginning of the two-year waiting period. Perelman himself seems oblivious to all such prognostications.

But let us assume for the moment that the question of publication has been settled and the proof has been accepted as valid. How should the prize money be allocated? I pointed out in the previous chapter that it would be absurd to credit all previous contributors to a subject since this would often lead all the way back to Newton and Leibniz. But in the case of the proof of Poincaré's Conjecture, the situation is clear. Hamilton's use of the Ricci flow was clearly instrumental. Without it, Perelman would have had nothing to work on.

Beyond that, the picture is a bit confused. Could one ever quantify the contributions of all the mathematicians who played a role in the proof of the Poincaré Conjecture? Practically everybody is in agreement about the following facts: Hamilton did the lion's share of the preliminary work, and Perelman scaled the last, and most difficult, mountain. But this is not the whole story.

During the past century, hundreds of mathematicians contributed in one way or another to the proof of the Poincaré Conjecture, starting with Poincaré himself and the early topologists, to those who reduced the question from topology to a manifold question, to those who adapted it to geometry. And what of all the analysts who laid the framework of the theory of differential equations? Hamilton even uses a technique first applied by John Nash that actually appeared on a blackboard in the background of the movie *A Beautiful Mind*.

What of the many mathematicians mentioned in Perelman's bibliographies: Altschuler, Anderson, Bando, Bakry, Cao, Cheeger, Chow, Colding, D'Hoker, Ecker, Gage, Gawedzki, Grayson, Gromov, Gross, Hildebrandt, Huisken, Ivey, Lawson, Li, Morrey, Waldenhausen? What of all the foundational efforts of those who first studied the heat equation, or minimal surface theory, or the work that contributed to those works? And what about those whose work is needed for the greater picture that Perelman developed, which will lead above and beyond even three dimensions?

Morgan and Tian mention more contributors, including many who worked to complete the details, who worked on the verification, and whose work was applied to the works of others. They add Burago, Minicozzi, Greene, Hempel, Jost, Ladyzhenskaya, Tam, Sachs, Shi, Shioya, Solonnikov, Stallings, Wu, Ural'oeva, Uhlenbeck, Yamaguchi, among others, and still do not do more than just scratch the surface.

There is one question, however, that the directors of the Clay Mathematics Institute, the scientific advisory board, and everybody else never considered. What if the prizewinner doesn't want the money? In our materialistic, money- and success-oriented world, the possibility of such a turn of events had entered nobody's mind. Yet this may be exactly what happens. Perelman, who rejected the Fields Medal, is likely to spurn the Millennium Prize too.

For the moment, it is still anybody's guess what will happen when the directors of the Clay Mathematics Institute award the million dollars. In his interview with *The New Yorker*, Perelman did not commit himself. Most people who know him believe that he cares as little about money as he does about the Fields Medal. Some pundits claim that he is afraid of becoming a millionaire. Remember how he was mugged in Berkeley when he carried only a few dollar bills on him? How much more dangerous it would be to carry thousands of rubles around in Russia. Whatever the reasons, Perelman is again keeping up the suspense. If we did not know better, we could believe that he was a masterful spin doctor. After all, receiving a million bucks is cool. Spurning a million bucks is just so much cooler.

In an interview at the ICM in Madrid, Carlson said that the Fields Committee "offered the medal based on Perelman's achievements and did not condition the decision on his possible reaction. If the Clay Institute decides to offer the Millennium Prize to Perelman, it will follow the same philosophy."

With this, I come to the end of the story. In fact, it is an era that comes to an end. A brilliant man's conjecture about the nature of space has been proven, and the only remaining question is when the million-dollar prize will be awarded…truly an insignificant detail in the pursuit of human knowledge.

But mathematics never ends. The successful solution to a problem just opens the doors to a host of new ones. It is easy to remain humble in the face of mathematics; so many problems have not been solved. Let's allow the magnificent enterprise to continue.

Notes

CHAPTER 2

10: on a flat disk

To illustrate the mathematical concept of a torus, I use bagels instead of doughnuts because doughnuts in England are actually jelly-filled *balls*.

11: are three-dimensional

Before embarking on the history of the Poincaré Conjecture, I made a promise. In the spirit of the German mathematician David Hilbert (1862–1943), Poincaré's contemporary who claimed that one should be able to explain mathematical findings "to the first man whom you meet on the street," I vowed that this book would contain no mathematical equations. But for the mathematically literate, it does help to frame a fact in mathematical notation. After all, that is why it was developed. So I will make an exception in this case. The ball is described by the equation $x^2 + y^2 + z^2 \leq 1$, while the surface of a ball is described by $x^2 + y^2 + z^2 = 1$. (The only other exception to my vow is in chapter 5.)

11: must be added

Unfortunately, "height above sea level" is not unambiguous. When the Germans and the Swiss built a bridge over the river Rhine in 2003, they discovered a height difference of fifty-four centimeters between the two sides. The Germans had used the North Sea, the Swiss the Mediterranean, as a baseline.

12: *just by deformations*

An interesting mathematical question—quite unrelated to topology—is how one can turn a ball into two balls of the same volume. The construction that involves judiciously cutting up the ball and rearranging the pieces is known as the Banach-Tarsky Paradox. For more on this paradox see Leonard Wapner's *The Pea and the Sun*.

13: *let alone proved*

Strangely, the answers are known for dimensions eight and twenty-four. For more on this subject, see my book *Kepler's Conjecture: How some of the greatest minds in history helped solve one of the oldest math problems in the world* (New York: John Wiley, 2003). Warning: Kepler's Conjecture is also known as the problem of the densest packing of *spheres* in three dimensions. But geometricians often talk about *spheres* when they actually mean *balls*. Topologists get rather bent out of shape about the distinction.

CHAPTER 3

16: *at London University*

When Poincaré died, he had the distinction of having received the most nominations for a Nobel Prize (for physics) of any nonwinner.

23: *physical fitness, and…drawing*

Today's curriculum at the École des Mines also includes a heavy dose of economics and management courses. Graduates no longer serve just as civil servants but are active in private industry.

23: *to the advanced program of the* Corps des Mines

A graduate of the École Polytechnique is called a *polytechnicien,* or *ancien élève de l'École Polytechnique* (former student of the École Polytechnique).

25: *a graduate of the* École des Mines

His final grade, 1,672 points out of a possible total of 2,100, was a respectable achievement for a French student indeed. (For the grand totals, the school did round the numbers to the nearest integers.)

CHAPTER 4

34: *three daughters and a son*

Jeanne in 1887, Yvonne in 1889, Henriette in 1891, and Léon in 1893.

35: *data available at the time*

For more on Kepler and Brahe, see my book *Kepler's Conjecture.*

37: *twelve entries had been received*

A thirteenth entry sent to the king directly was not accepted for the competition because it arrived six months after the deadline. That was just as well, as its author, one Cyrus Legg from Clapham, England, was a well-known quack.

37: 'The problem of the three bodies'"

Actually, the full name of Poincaré's entry was *Sur le problème des trois corps et les equations de la dynamique,* but Hollywood prefers short, concise titles.

37: *the king signed the protocol*

Even though the name of the essay's author had officially been a secret until this moment, the members of the jury had been aware of the winner's identity all along since Poincaré and Mittag-Leffler had kept up a lively correspondence throughout the past years.

40: the system could explode

Even if the system in actual fact *is* stable, Gyldén would not have proved that with the help of an approximation that does not converge.

46: take a very, very long time

The recurrence theorem provides a formidable challenge to the second law of thermodynamics, which states that entropy always increases. The theorem and the law can be reconciled by observing that external circumstances change over a very, very long time and the system is thus not left to itself.

51: three months after Henri died

Émile's son, Pierre Boutroux, also became a mathematician of note. Educated at École Normale in Paris, he taught at Princeton—even becoming the chairman of its graduate mathematics department—but returned to France to fight bravely in the First World War and died at age forty-two in 1922.

CHAPTER 5

54: would cross all the bridges

Many sources claim that Ehler was the mayor of Danzig when he had his correspondence with Euler. This is not quite true. Ehler was mayor for three terms (1741–42, 1745–46, and 1750–51), none of which coincided with the period under discussion.

57: reported a remarkable fact

Yes, the one from the famous Goldbach Conjecture ("Every even number can be expressed as the sum of two primes").

57: one always gets 2

In mathematical notation, $v - e + f = 2$, where v, e, and f stand for vertices, edges, and faces.

58: that gives us 4, not 2

Hence $v - e + f$ is 4.

59: takes into account hidden cavities

His appended equation reads $v - e + f = 2 + 2p$, where p is the number of hidden cavities contained in the body.

60: everything we are about to say still holds

The complete formula reads $v - e + f = 2 - 2g + 2p + c$, where g is the number of tunnels and c stands for the number of faces that have bulges or dents or entrances to tunnels. In the absence of holes, bulges, and cavities, l'Huilier's formula reduces to Euler's formula for solids. The number of tunnels, g, is called the body's genus.

62: six others, refused

Among them were also Gauss's son-in law, Georg Ewald, and the brothers Grimm.

63: just the right subject

Then again, Gauss never published anything on non-Euclidean geometry either, which would later become an extremely important subject.

65: of course, the Möbius strip

The Möbius strip can easily be modeled as a strip of paper twisted once and glued to itself.

67: *"until the fruits of his mind matured."*

R. Baltzer, F. Klein, and W. Schiebner, eds., *August Möbius, Gesammelte Werke* (Leipzig: 1885–87).

68: *than it has dimensions*

Actually it has more, but the others are all zero.

68: *because they have a hole*

The technical term for this shape is *annulus*.

72: *inelegantly named HOMFLY polynomial*

HOMFLY is an acronym for its discoverers: Hoste, Ocneanu, Millett, Freyd, Lickorish, and Yetter.

74: *for this young discipline*

In this brief overview of the history of topology, many important players have been left out. Let me mention just two of them. Bernhard Riemann (1826–1866) was one of the mathematicians whom Betti had befriended on his 1858 trip to Göttingen. This most brilliant of Gauss's students is well-known today even among many nonmathematicians because of the still unsolved Riemann Conjecture. Camille Jordan (1838–1922) proved a pathbreaking theorem that says that a closed curve, such as a circle, divides a plane into exactly two regions: an inside and an outside. The reference to "pathbreaking" is only partly tongue-in-cheek because the theorem's simplicity is deceptive. It was Jordan's contribution to realize that a rigorous proof for this seemingly obvious result was necessary.

CHAPTER 6

78: *counterexample in four-dimensional space*

Poincaré constructed a pair of four-dimensional spaces that have the same Betti numbers but are topologically distinct.

79: "So I preferred to be a bit talkative"

This calls to mind the apology of an Israeli Supreme Court judge in the preface to a particularly long opinion: "I did not have enough time to be brief."

79: calling on his senses

Les figures suppléent d'abord à l'infirmité de notre ésprit en appelant nos sens à son secours. Actually, Poincaré spoke of functions of two complex variables, which need to be imagined in four-dimensional space.

81: object's k-dimensional connectivity

k runs from zero to n. The manifolds must also be *orientable*, which means that manifolds of the Möbius type are excluded. The technical term *closed* means "compact and without boundary."

81: floating in four-dimensional space (1,3,3,1)

Recall that for the one-dimensional circle there is only one piece, so $b_0 = 1$, and one hole, so $b_1 = 1$. Hence, b_0 equals b_1 just as the duality theorem predicts. For the two-dimensional sphere there is one piece, so $b_0 = 1$, there are no holes, so $b_1 = 0$ and there is one chamber inside, so $b_2 = 1$. We have b_0 equals b_2. Similarly a bagel has one piece, $b_0 = 1$, two holes (one in the center, and one tunneling through the inside), so $b_1 = 2$, and one chamber, so $b_2 = 1$ and, again as predicted, b_0 equals b_2.

82: in terms of mathematics

In this he had the exact same experience as his fellow Scandinavian Niels Henrik Abel nearly seventy years earlier.

86: for reduced Betti numbers

The revised version of the duality theorem, now called Poincaré Duality, is frequently used by mathematicians to this day.

89: *(the visual substrate)*

Note that the German word *Anschaulichkeit* on which *Anschauungssubstrat* is based is not accurately translatable. It means something like visually intuitive or descriptive.

89: *favored and recommended*

This "intuitive descriptiveness" would later become one of the hallmarks of the *"Deutsche Mathematik"* so favored by the Nazis, in distinction to the allegedly abstract "Jewish mathematics." Of course, Felix Klein had nothing to do with the later aberrations.

91: *oh, yes, the Poincaré Conjecture*

The four-color problem was solved only in 1976 with the help of a massive computer effort.

94: *on a firm foundation*

Actually, he should have said *conjecture,* since it was not a theorem yet.

CHAPTER 7

95: *topologically equivalent shape*

The even more technical term for "topologically equivalent" is *homeomorphic.*

96: *convex, three-dimensional body*

I.e., the solution to the equation remains the same even if the body is squeezed and squashed.

96: *these numbers are topologically different*

Unfortunately, the converse is not true: Bodies with identical homology groups may still be topologically different.

97: "sphere is homeomorphic to it"

So even though bodies with identical homology groups may, in general, be topologically different, Poincaré thought that this was an exceptional case.

97: three and a half centuries

Only in 1993 was Andrew Wiles from Princeton University able to prove that the equation $x^n + y^n = z^n$ has no solutions for integers n greater than two.

99: the surface of the ball is to a disk

And in the same manner that disks are slices of the ball, balls are slices of the four-dimensional object bounded by the three-dimensional balls.

99: shadow does not reveal everything

This was already pointed out by Plato in his famous "Allegory of the Cave."

99: lists eight different methods

R. C. Kirby and M. G. Scharlemann. See bibliography.

100: bounded by twelve regular pentagons

It is one of the five regular platonic solids, the other ones being the cube, the tetrahedron (the pyramid with triangular base), the octahedron (two tetrahedra glued together at the bases), and the icosahedron (a twenty-faced polyhedron).

101: second Betti number equal to one

Another way to imagine Poincaré's sphere is to regard it as the quotient space SO(3)/I where I is the symmetry group of the regular icosahedron. This means that the Poincaré sphere is the space of all possible positions of an icosahedron.

106: *quod erat demonstrandum*

Or QED. "This is what had to be proved."

107: *illustrate this in two dimensions*

They are important for the higher-dimensional version of the Poincaré Conjecture.

108: *mapped to the parachutist's backpack*

This includes the space between the lines. Better still, imagine the canopy extending all the way to the backpack.

108: *they are called equivalent*

Don't ever try that with real parachutes. As soon as two parachutes overlap, one of them collapses.

109: *cannot be morphed into each other*

In more technical language, fundamental groups of homeomorphic spaces are isomorphic.

109: *must they be topologically equivalent?*

The group need not be the same but must be isomorphic.

110: *Poincaré actually thought so.*

While Poincaré's homology sphere had no holes that could be identified using Betti numbers, he proved it had a highly nontrivial fundamental group. Remember that this space is created from a dodecahedron with opposite sides identified with a twist. So an interesting loop in this space, which cannot be stretched or shrunk to a point, starts at the center of the dodecahedron, runs to the center of one of the pentagonal faces, at which point it is glued to the center of the opposite pentagonal face and runs straight back to the center. So this space does not have a trivial fundamental group.

110: not be homeomorphic to a sphere?

Poincaré's original question reads, *"Est-il possible que le groupe fondamental de V se réduise à la substitution identique, et pourtant V ne soit pas simplement connexe?"* (*V* stands for a manifold.) Literally this would translate as "Is it possible that the fundamental group of *V* reduce to the identity element [i.e., the group is trivial] and *V* nevertheless not be simply connected?" This formulation gave rise to some confusion because *trivial fundamental group* and *simply connected* are today used synonymously. Hence the question sounds suspiciously false to modern readers. John Milnor cleared up the problem in 2003 in a footnote to a paper describing the status of the conjecture: "Poincaré's terminology may confuse modern readers who use the phrase 'simply-connected' to refer to a space with trivial fundamental group. In fact, he used 'simply-connected' to mean homeomorphic to the simplest possible model, that is, to the three-sphere."

111: subject classification number for Poincaré's Conjecture

57M40: "Characterizations of E^3 and S^3 (Poincaré conjecture)."

CHAPTER 8

114: north of Oxford

Which may have been, but probably was not, the model for Manor Farm in George Orwell's *Animal Farm*.

114: price had become too high

For the year 2007, the subscription price for institutions was $1,665 for six issues.

118: in Munich at the time

Carathéodory was a professor at the University of Munich until 1938 and continued to live and work in Germany throughout World War II—making doubtful arrangements with the Nazis and Nazi sympathizers.

120: *embedded into three-dimensional manifolds*

One of the theorems was strengthened somewhat by Henry White-head.

123: *to avoid any such confusion*

Maskit's counterexamples and Papa's response were communicated to the *Bulletin* on January 29, 1963.

124: *a former student of Max Dehn's*

Magnus, who had fled Nazi Germany, was quite a Ph.D.-producing machine, turning out an average of three Ph.D.'s a year during his quarter century at NYU and at Polytechnic University.

134: *"a combinatorial approach"*

The combinatorial approach consists of investigating all possible combinations of gluing tetrahedra together.

138: *string is knotted or not*

Haken had to assume that the string was piecewise linear, which means that it is made of a collection of line segments meeting at corners, like a carpenter's measuring rod. His proof used Papa's proof of Dehn's Lemma.

139: *become standard fare since then*

Kepler's Conjecture was proved by brute force. (See my book *Kepler's Conjecture*.)

CHAPTER 9

146: *twisted within the plane*

The video *Outside In* produced at the Geometry Center (formerly at the University of Minnesota) in 1994 is an animated film of sphere eversion and explains the issue with train tracks.

149: do their work anywhere

Gauss once remarked that he did his best mathematical work in bed. More recently I once observed Tim Gowers, a Fields Medalist from Cambridge University, in an airport on a small island in Greece. His plane back to England had been delayed for a few hours. While most passengers were hopping mad, Gowers did not seem to mind at all. He just pulled out a pad and started working on a mathematical problem right there, in the stuffy departure hall.

149: for gender discrimination

The suit ended with a settlement.

151: The Smale Collection: Beauty in Natural Crystals

Alas, the leatherbound copy, which went for $250, is no longer available.

153: the world's leading topologists

So highly regarded was Eilenberg that a letter from the Soviet Union, addressed to "Mathematics, USA," was delivered to him by the U.S. Postal Service. By the way, what minerals are to Smale, Asian sculptures were to Eilenberg. He spent decades building up one of the world's most important collections of small-scale Asian sculptures and became known as the foremost authority on it. He bequeathed his collection, valued at $5 million, to the Metropolitan Museum of Art.

155: suggestions from the floor

Now the Arbeitstagungen are held biannually.

157: had probably saved his life

One may be tempted to say that the atomic bomb saved his life with a probability of 60 percent.

162: famous Wallace-Lickorish Theorem
See chapter 8.

163: in the Journal of Mathematics and Mechanics
This journal would later become *The Indiana University Mathematics Journal.* "Modifications and Cobounding Manifolds I" had appeared a year earlier in the *Canadian Journal of Mathematics.*

167: whether P equals NP
A difficult question about complexity in computer science, which we shall not go into. See, for example, David Harel, "Computers, Ltd.: What they really can't do."

171: the four-dimensional sphere is one of them
Freedman's classification does not cover the case of nonsimply connected four-dimensional manifolds. Also, while his work gives the homeomorphism classification, the diffeomorphism classification (smooth as opposed to merely continuous objects) is still wide-open.

CHAPTER 10

174: in Palo Alto, California
This institute, founded and financed by the owners of Fry Electronics, is hidden right next to the flagship store near the campus of Stanford University. This author, not knowing where to look, wandered in and out of Fry's wonderful store, which carries just about everything a computer geek craves, including potato chips and Mars bars. Finally he found the small side door next to the store that led to AIM. I would like to thank AIM for making a copy of Armentrout's manuscript available to me.

175: let us call them alpha-spaces
To be precise, it dealt with "homogenous compact ANR-spaces," where ANR stands for "absolute neighborhood retract."

177: the latter is called collapsing

I earlier talked about "collapsing" and "contracting" in chapter 8.

179: before it had even begun

See chapter 8 of my book *The Secret Life of Numbers*.

CHAPTER 11

186: known as Thurston's Geometrization Conjecture

Note that Thurston is a geometric analyst and always worked with smooth manifolds that have no corners.

186: appointed full professor at Princeton University

His Ph.D. thesis was entitled "Foliations of Three-Manifolds which are Circle Bundles." A circle bundle is a manifold made by gluing infinitely many circles together like a cylinder (where the circles are glued along a line) or a torus (where the circles are glued around a ring). A foliation is a way of slicing a manifold into leaves so that it can be riffled through like the pages of a book. Remember the Euler characteristic we defined in chapter 5, which involved adding up vertices and faces and subtracting edges? One of Thurston's best-known results was that any manifold with Euler characteristic zero has a codimension one foliation. The mathematics world was quite surprised that this number could be used to show that a manifold of any dimension could be sliced into infinitely many sheets.

188: divided only by one and by themselves

In chapter 5 we also saw that so-called prime knots form the building blocks for even the most complicated knots.

188: can be decomposed into primafolds

Only in 1962, however, did John Milnor prove that the decomposition of every manifold into its primafolds is unique. In 1998, Jaco and

Rubinstein designed an algorithm that specified a manifold's prima-folds.

188: *of which all manifolds are made*

Recall that the number 1 is not considered a prime number. Similarly its counterpart, the sphere, is not considered a prime manifold.

188: *are henceforth ignored*

This surgical operation is called forming a connected sum.

190: *raised by an additional dimension*

If you would like to go flying through three-dimensional hyperbolic space, take a look at Jeff Weeks's "Curved Space." Weeks is an independent mathematician who won a MacArthur genius award. His adviser was Thurston.

191: *one direction shows a flat geometry*

The first two shapes can also be regarded as bundles, similar to the circle bundles studied in Thurston's thesis. Recall that we said a cylinder is a bundle of infinitely many circles glued along a line. One can also create a sphere bundle by gluing infinitely many spheres along a line, and a hyperbolic plane bundle by gluing infinitely many hyperbolic planes along a line. One might also glue infinitely many Euclidean planes along a line, but that just gives Euclidean three-space, so we have already included it.

191: *two-dimensional building blocks*

Nil is a circle bundle over a torus with an extra twist to it. Sol is a torus bundle over a circle with an extra twist. Strangely, if one takes a circle bundle over a sphere with an extra twist, one gets another sphere! Try to wrap your mind around that.

192: a "steady string of girlfriends"
The New Yorker, August 22, 2006.

193: in partial differential equations
The third recipient of that year's Fields Medal was Alain Connes.

197: to a geometrical phenomenon
Albert Einstein used tensors and the Ricci curvature to design an equation that imitates the undulation of waves, thus producing the Einstein equation used in general relativity to describe gravity.

197: to tensor calculus
Of course, Einstein wanted to understand tensor calculus on a deep level rather than just use it by rote.

197: soon his innards get warm
On the other hand, if no vodka or tea is forthcoming, his body eventually goes cold, first the fingers, then the arms, until...well, read Jack London's "To Build a Fire" for an account of the thermodynamics of heat dissipation within a human body.

197: fits the curve at this point
Osculating derives from the Latin word for *kissing* and indicates something a little more tender than *tangent* (which comes from the Latin for *touching*).

198: the second in the z-direction
This is assuming the curve is traveling in the *y*-direction.

198: or higher-dimensional space
For simplicity, I have described curvature "extrinsically," as viewed from the outside. Ricci-curvature is in fact measured from inside the manifold.

199: for distances on the manifold

Actually for the rate of change of the distances.

200: rapidly shrinking manifold of constant curvature

This is similar to heat that diffuses until the whole body is of uniform temperature. But in contrast to heat diffusion, where hot regions cool down and cool regions heat up, the Ricci flow does the opposite: Relatively flat regions curl up slowly, curved regions curl up even more.

200: tries to bend itself into shape

Instead of straightening out a manifold, one can adjust the scale. In differential geometry this amounts to the same thing. One can shorten a distance or one can enlarge the scale.

201: with the tools of another (differential equations)

One major step in Hamilton's proof was to show that given any smooth manifold, there exists a Ricci flow. So it is necessary that the manifold be smooth. Luckily, Bing and Moise (see chapter 8) had proven that any three-dimensional manifold can be triangulated and thus is homeomorphic to a smooth manifold. Otherwise Hamilton's technique would not be able to prove the Poincaré Conjecture.

203: suspended in four-dimensional space

By analogy, think of a partly curled-up sheet of paper—a two-dimensional object suspended in three-dimensional space—as the product of two one-dimensional objects: a parabola multiplied by a line. A parabola rotated through 360 degrees, instead of pushed along a line, would produce an object like a satellite dish. This is also a two-dimensional object, suspended in three dimensions. On the other hand, one could also think of this object as a bundle of infinitely many cigars glued along a line.

CHAPTER 12

210: *spheres and long, thin tubes*

Actually, a whole taxonomy of singularities developed, including—apart from tubes—circuits, caps, and horns. We will not deal with all of the subspecies.

211: *Thus the Poincaré Conjecture holds*

Remember the Poincaré Conjecture says that a closed, three-dimensional manifold with a trivial fundamental group must be a three-dimensional sphere.

212: *while focusing on a developing singularity*

Parabolic rescaling is used in such diverse fields as the study of subsonic airflow and stochastic probability.

213: *"sometimes a cigar is just a cigar"*

This was in the context of a discussion of phallic symbols.

214: *infinitely often for endless time*

Another approach, the so-called thick-thin decomposition, leads to a proof of the Geometrization Conjecture.

214: *that span minimal areas*

Mimimal surface is a technical term by which mathematicians mean the smallest possible surface spanning the frame.

215: *sufficient skin to grow new bodies*

Actually, it is not really the area of the skin that is finite but a technically complicated surface area that is related to minimal surfaces or soap bubbles.

215: *thus proving Poincaré's Conjecture*

Note that this estimating of areas requires the manifold to be simply connected, so with this method one proves only Poincaré's Conjecture, not geometrization. Colding and Minicozzi provided an alternate proof of this step of the problem. Their proof is based on something called the width of the manifold and also uses minimal-surface theory.

216: *have to cope with the fallout*

Recall the Oxenhielm incident on page 179.

223: *invited Perelman to give these lectures*

The approach Anderson had studied was one outlined by Thurston in which each manifold is assigned a special "best shape." The sphere's best shape is round rather than ellipsoidal. This best shape is found by studying the total scalar curvature—which is like an average of the Ricci curvature—and looking for critical points. For those of you who have taken calculus, these are the same critical points that one looks for in calculus, only the calculus is being performed on an infinite dimensional space. The approach was motivated by the Euler characteristic, which had been so useful for characterizing surfaces and its relationship to the average curvature of surfaces: the so-called Euler formula. (See chapter 5.)

CHAPTER 13

226: *from the University of Michigan*

Kleiner is now at Yale.

238: *a very public and very bad black mark*

The events parallel those of a few years earlier when the Chinese-born professor of mathematics Wu-Yi Hsiang rushed through a hundred-page faulty paper on the Kepler Conjecture in the *International Journal of Mathematics*. The editors were friends of his at the Berkeley math department, and it took them just four months to vet the erroneous proof. For an account of these events see my book *Kepler's Conjecture: How some of the greatest minds in history helped solve one of the oldest math problems in the world* (John Wiley, 2003).

240: 2.5 percent of the gross domestic product

The USA spends about 2 percent of GDP on research.

241: standing on the shoulders of Giants

This prompted a wag to say, "If I haven't seen as far, it is because giants are standing on my shoulders."

244: bad blood could have been avoided

I have tried contacting Huai-Dong Cao to get his views but unfortunately never received any answer. My request to Shing-Tung Yau for further information was answered with a detailed six-page e-mail in which Yau set out the points with which he did not agree. Unfortunately, he did not authorize me to quote from his letter. Thus, regretfully, Yau's views may not be fully reflected in this book. I have nevertheless tried to give an accurate account of the events as I see them.

CHAPTER 14

247: jointly with his brother Johann II

The elder Bernoulli would never forgive his son for winning part of the prize. See chapter 4 of my book *The Secret Life of Numbers*.

248: the French Academy of Sciences, in 1888

She did not attend the awards ceremony because she believed the scientific community did not show her appropriate respect.

251: taken in by grave robbers

For the sanctimonious among us, I suggest taking a walk through chaotic Cairo Museum and asking ourselves whether the ancient Egyptian artifacts are not displayed in a more seemly manner in the British

Museum. Were it not for the "looters," many artifacts would probably have served as building materials.

254: *are identical to the Millennium Problems*

One of them that does not appear on the Millennium list is "What are the limits of intelligence, both artificial and human?"

254: Uncle Petros & Goldbach's Conjecture *by Apostolos Doxiadis*

The Goldbach conjecture says that every even integer greater than 2 is the sum of two primes (e.g., $8 = 5+3$).

Bibliography

The items listed in the bibliography are an eclectic assortment of books and articles on the Poincaré Conjecture. The list is not necessarily meant as a teaser for "further reading" nor is it exhaustive. It is intended to give readers interested in the historical or technical aspects of the matters that are discussed in the book an indication of where to start. Many items are technical in nature and directed toward specialists.

BOOKS

Baltzer, R., F. Klein, and W. Schiebner, eds. *August Möbius, Gesammelte Werke*. Leipzig, 1885–87.

Banyaga, A., H. Movahedi-Lankarani, and R. Wells, eds. *Topics in Low-Dimensional Topology (In Honor of Steve Armentrout)*. World Scientific Publishing, 1999.

Barrow-Green, June. *Poincaré and the Three-Body Problem*. American Mathematical Society, 1996.

Bell, E. T. *Men of Mathematics*. New York: Simon & Schuster, 1937.

Carlson, James, Arthur Jaffe, and Andrew Wiles, eds. *The Millennium Prize Problems*. Providence, RI: American Mathematical Society, 2006.

Chasles, M. *Aperçu Historique sur l'Origine et le Développement des Méthodes en Géométrie.* Paris: Gauthier-Villars, 1889.

Chow, Bennett, and Dan Knopf. *The Ricci Flow: An Introduction.* Providence, RI: American Mathematical Society, 2004.

D'Adhemar, Vicomte Robert. *Henri Poincaré.* Paris: A. Hermann & fils, 1912.

Denjoy, Arnaud. *Hommes, Formes et le Nombre.* Paris: Albert Blanchard, 1964.

Devlin, Keith. *The Millennium Problems.* New York: Basic Books, 2002.

Dieudonné, Jean. *History of Algebraic and Differential Topology, 1900–1960.* Basel: Birkhäuser, 1989.

Doxiadis, Apostolos. *Uncle Petros & Goldbach's Conjecture.* New York: Bloomsbury, 2001.

Fort, M. K. *Topology of 3-Manifolds and Related Topics, Proceedings of the University of Georgia Institute.* New Jersey: Prentice-Hall, 1962.

Gardner, Martin. *The Sixth Book of Mathematical Games from Scientific American.* Chicago: University of Chicago Press, 1984.

Halmos, Paul R. *I Want to Be a Mathematician.* Berlin: Springer, 1985.

Harel, David. *Computers, Ltd.: What They Really Can't Do.* Oxford: Oxford University Press, 2003.

Hatcher, Allen. *Algebraic Topology.* Cambridge: Cambridge University Press, 2002.

Hirsch, M. W., J. E. Marsden, and M. Shub, eds. *From Topology to Computation: Proceedings of the Smalefest.* Berlin: Springer Verlag, 1993.

James, I. M., ed. *History of Topology.* North Holland, 1999.

Le Livre du Centenaire de la Naissance de Henri Poincaré. Paris: Gauthier-Villars, 1955.

Picard, Émile. *L'Oeuvre d'Henri Poincaré.* Paris: Extrait de la Revue Scientifique, 1913.

Poincaré, Henri. *Sur le problème des trois corps et les équations de la dynamique.* Mémoire couronné de prix de sa Majesté le Roi Oscar II, 1889.

Pont, Jean-Claude. *La Topologie Algébrique des origins à Poincaré.* Paris: Presse Universitaire de France, 1974.

Scholz, Erhard. *Geschichte des Mannigfaltigkeitsbegriffes von Riemann bis Poincaré.* Basel: Birkhäuser, 1999.

Scriba, Christoph J., and Peter Schreiber. *5000 Jahre Geometrie.* Berlin: Springer Verlag, 2001.

Segal, Sanford. *Mathematicians Under the Nazis.* Princeton: Princeton University Press, 2003.

Smale, Steve. *The Smale Collection: Beauty in Natural Crystals.* East Hampton, CT: Lithographie LLC, 2006.

Szpiro, George G. *Kepler's Conjecture: How some of the greatest minds in history helped solve one of the oldest math problems in the world.* New York: John Wiley, 2003.

———. *The Secret Life of Numbers: 50 Easy Pieces on How Mathematicians Work and Think.* Washington, D.C.: Joseph Henry Press, 2006.

Topping, Peter. "Lectures on the Ricci Flow." Unpublished.

Wapner, Leonard. *The Pea and the Sun.* Wellesley, MA: A. K. Peters Ltd., 2005.

ARTICLES

Anderson, Michael T. "Geometrization of 3-manifolds via the Ricci flow." *Notices of the AMS* 51 (2004): 184–93.

Bessières, L. "Poincaré Conjecture and Ricci flow: An outline of the work of R. Hamilton and G. Perelman." *EMS Newsletter* (2006): 11–22.

Bing, RH. "Some aspects of the topology of 3-manifolds related to the Poincaré Conjecture." In T. L. Saaty, ed., *Lectures on Modern Mathematics*, 2:93–127. New York: Wiley, 1963–65.

———. "The Kline sphere characterization problem." *Bulletin of the AMS* 52 (1946): 644–53.

———. "Necessary and sufficient conditions that a 3-manifold be S^n." *Annals of Mathematics* 68 (1958): 17–37.

———. "An alternative proof that 3-manifolds can be triangulated." *Annals of Mathematics* 69 (1959): 37–65.

Burde, Gerhard, Wolfgang Schwarz, and Jürgen Wolfart. "Max Dehn und das mathematische Seminar." Frankfurt am Main, 2002.

Cao, Huai-Dong, and Xi-Ping Zhu. "A complete proof of the Poincaré and geometrization conjectures—Application of the Hamilton-Perelman theory of the Ricci flow." *Asian Journal of Mathematics* 10 (2006): 165–492.

———. "Erratum to 'A complete proof of the Poincaré and geometrization conjectures—Application of the Hamilton-Perelman theory of the Ricci flow.'" *Asian Journal of Mathematics* 10 (2006): 663–64.

Ciesielski, Krzysztof, and Zdzisław Pogoda. "Interview with Ian Stewart." *EMS Newsletter* (2005): 26–33.

Colding, Toby, and Bill Minicozzi. "Estimates for the extinction time for the Ricci flow on certain 3-manifolds and a question of Perelman." *Journal of the AMS* 18 (2005): 561–69.

Darboux, Gaston. "Éloge historique d'Henri Poincaré." *Mémoires de l'Académie des Sciences* 52 (1913).

Dawson, John W., Jr. "Max Dehn, Kurt Gödel and the Trans-Siberian Escape Route." *Notices of the AMS* 49, no. 9 (2002): 1068–75.

Dehn, Max, and Poul Heegaard. "Analysis situs." In *Enzyklopädie der mathematischen Wissenschaften*, vol. 3, ed. W. F. Meyer and H. Mohrmann. Leipzig: B. G. Teubner, 1923.

Ewing, J. H., W. H. Gustafson, Paul R. Halmos, et al. "American mathematics from 1940 to the day before yesterday." *American Mathematical Monthly* 83 (1976): 503–16.

Freedman, Michael Hartley. "The topology of four-dimensional manifolds." *Journal of Differential Geometry* 17 (1982): 357–453.

Gilman, D., and Dale Rolfsen. "The Zeeman conjecture for standard spines equivalent to the Poincaré Conjecture." *Topology* 22 (1983): 315–23.

Heegaard, Poul. "Sur l'analysis situs." *Bulletin de la Société Mathématique Française* 44 (1916): 161–242.

Hilton, Peter J. "Memorial tribute to J. H. C. Whitehead." *L'enseignement mathématique* 7 (1961): 107–24.

Jackson, Allyn. "The Clay Mathematics Institute." *Notices of the AMS* 46 (2006): 888–89.

———. "Conjectures no more?" *Notices of the AMS* 53 (2006): 897–901.

Jakobsche, W. "The Bing-Borsuk conjecture is stronger than the Poincaré Conjecture." *Fundamenta Mathematicae* 106 (1980): 127–34.

James, I. M. "Portrait of Alexander (1888–1971)." *Bulletin of the AMS* 38 (2001): 123–29.

Kelley, Paul. "Report of the memorial resolution committee for RH Bing." *Documents and Minutes of the General Faculty,* University of Texas at Austin, undated.

Kirby, R. C., and M. G. Scharlemann. "Eight faces of the Poincaré homology 3-sphere." In J. C. Cantrell, ed., *Geometric Topology, Proceedings of the 1977 Georgia Topology Conference,* 113–46. Academic Press, 1979.

Kleiner, Bruce, and John Lott. "Notes on Perelman's Papers." arXiv: math.DG/0605667 (2006).

Koseki, Keniti. "Bemerkungen zu meiner Arbeit 'Poincarésche Vermutung.'" *Mathematical Journal of Okayama University* 8 (1958): 1–106.

————. "Bemerkungen zu meiner Arbeit 'Poincarésche Vermutung.'" *Mathematical Journal of Okayama University* 9 (1960): 165–72.

Maskit, Bernard. "On a conjecture concerning planar coverings of sufaces." *AMS Bulletin* 69 (1963): 306.

Mawhin, Jean. "Henri Poincaré: A life in the service of science." *Notices of the AMS* 52 (2005): 1036–44.

Milnor, John W. "The Work of J. H. C. Whitehead." In *J. H. C. Whitehead, Collected Works,* vol. 1. New York: Pergamon Press, 1962.

————. "The Poincaré Conjecture 99 years later: A progress report." Unpublished preprint.

Moise, Edwin E. "Affine structures in 3-manifolds. V. The triangulation theorem and Hauptvermutung." *Annals of Mathematics* 56 (1952): 96–114.

Morgan, John. "Recent progress on the Poincaré Conjecture and the classification of 3-manifolds." *Bulletin of the AMS* 42 (2004): 57–78.

Morgan, John W., and Gang Tian. "Ricci flow and the Poincaré Conjecture." arXiv: math.DG/0607607 (2006).

Munkholm, Ellen S., and Hans J. Munkholm. "Poul Heegaard." In *History of Topology*, ed. I. M. James. North Holland, 1999.

Nasar, Sylvia, and David Gruber. "Manifold Destiny." *The New Yorker,* August 28, 2006, 44–57.

Newman, M. H. A. "John Henry Constantine Whitehead, 1904–1960." *Biographical Memoirs of Fellows of the Royal Society* 7 (1961): 349–63.

Novikov, S. "Henri Poincaré and the XXth Century Topology." In *Proceedings of the Symposium Henri Poincaré.* Brussels: International Solvay Institute for Physics and Chemistry, 2004.

O'Connor, John, and Edmund Robertson. "MacTutor History of Mathematics." http://www-history.mcs.st-andrews.ac.uk/history/index.html.

Papakyriakopoulos, Christos D. "A reduction of the Poincaré Conjecture to other conjectures. II." *Bulletin of the AMS* 69 (1963): 399–401.

Papastavridis, Stavros G. "Christos Papakyriakopoulos: His Life and Work." Ethniko Metsovio Polytexneio, 2000.

Perelman, Grisha. "The entropy formula for the Ricci flow and its geometric applications." arXiv: math.DG/0211159 (2002).

———. "Ricci flow with surgery on three-manifolds." arXiv: math. DG/0303109 (2003).

———. "Finite extinction times for the solution to the Ricci flows on certain three-manifolds." arXiv: math.DG/0307245 (2003).

Poénaru, V. "A program for the Poincaré Conjecture and some of its ramifications." In *Topics in Low-Dimensional Topology* (*In Honor of Steve Armentrout*), ed. A. Banyaga, H. Movahedi-Lankarani, and R. Wells, 65–88. World Scientific Publishing, 1999.

Poincaré, Henri. "Analysis situs." *Journal de l'École Polytechnique* 1 (1895): 1–121.

————. "Sur les nombres de Betti." *Comptes Rendus de l'Académie des Sciences* 128 (1899): 629–30.

————. "Complément à l'analysis situs," *Rendiconti del Circolo Matematico di Palermo* 13 (1899): 285–343.

————. "Second complément à l'analysis situs," *Proceedings of the London Mathematical Society* 32 (1900): 277–308.

————. "Sur l'analysis situs." *Comptes Rendus de l'Académie des Sciences* 133 (1901): 707–9.

————. "Sur certaines surfaces algébriques; troisième complément à l'analysis situs." *Bulletin de la Société Mathématique de France* 30 (1902): 49–70.

————. "Les cycle des surfaces algébriques; quatrième complément à l'analysis situs." *Journal de Mathématique* 8 (1902): 169–214.

————. "Cinquième complément à l'analysis situs." *Rendiconti del Circolo Matematico di Palermo* 18 (1904): 375–407.

————. "Sur un théorème de Géométrie." *Rendiconti del Circolo Matematico di Palermo* 18 (1904): 45–110.

Repovs, Dusan. "The recognition problem for topological manifolds: A survey." *Kodai Mathematical Journal* 17 (1994): 538–48.

Roy, Maurice, and René Dugas. "Henri Poincaré, ingénieur des mines." *Annales des Mines* (1954).

Scott, Peter. "The geometries of 3-manifolds." *Bulletin of the London Mathematical Society* 15 (1983): 401–87.

Shalen, Peter B. "A 'piecewise-linear' method for triangulating 3-manifolds." *Advances in Mathematics* 52 (1984): 34–80.

Smale, Steve. "Generalized Poincaré Conjecture in Dimensions Greater Than Four." *Annals of Mathematics* 74 (1961): 391–406.

————. "A survey of some recent developments in differential topology." *AMS Bulletin* 69 (1963): 131–45.

————. "The story of the higher dimensional Poincaré Conjecture (What actually happened on the beaches of Rio)." *The Mathematical Intelligencer* 12 (1990).

————. "Finding a horseshoe on the beaches of Rio." *The Mathematical Intelligencer* 20 (1998).

Stallings, John. "Polyhedral Homotopy-Spheres." *AMS Bulletin* 66 (1960): 485–88.

————. "The piecewise-linear structure of Euclidean space." *Proceedings of the Cambridge Philosophical Society* (*Mathematics and Physical Sciences*) 58 (1962): 481–88.

————. "How not to prove the Poincaré Conjecture." Unpublished.

Strzelecki, Pawel. "The Poincaré Conjecture?" *American Mathematical Monthly* 113 (2006).

Taubes, Gary. "What Happens When Hubris Meets Nemesis." *Discover,* 1987.

Thickstun, T. L. "Open acyclic 3-manifolds, a loop theorem and the Poincaré Conjecture." *Bulletin of the AMS*, n.s., 4 (1981): 192–94.

Vershik, Anatoly. "What is good for mathematics? Thoughts on the Clay Millennium Prizes." *Notices of the AMS* 54 (2007): 45–47.

Wallace, Andrew H. "Modifications and cobounding manifolds." *Canadian Journal of Mathematics* 12 (1960): 503–10.

————. "Modifications and cobounding manifolds II." *Journal of Mathematics and Mechanics* 10 (1961): 773–809.

Whitehead, J. H. C. "Certain theorems about three-dimensional manifolds (I)." *Quarterly Journal of Mathematics* 5 (1934): 308–20.

————. "Three-dimensional manifolds (Corrigendum)." *Quarterly Journal of Mathematics* 6 (1935).

Yamasuge, Hiroshi. "On Poincaré Conjecture for M⁵." *Journal of Mathematics, Osaka City University* 12 (1961).

Zeeman, E. Christopher. "Unknotting spheres in five dimensions." *AMS Bulletin* 66 (1960): 198.

————. "The generalized Poincaré Conjecture." *AMS Bulletin* 67 (1961): 270.

Acknowledgments

M any people have offered information or read parts of the manuscript and generously made suggestions or corrected mathematical or historical errors: Colin Adams, Michael Anderson, Jim Arthur, Michael Barr, Laurent Bartholdi, Gilbert Baumslag, Maarten Bergvelt, Joan Birman, Christian Blatter, Juri Burago, Brian Conrey, Apostolos Doxiadis, Beno Eckmann, Greg Egan, Emmanuel Farjoun, Stefan Fredenhagen, Anandaswarup Gadde, Misha Gromov, Bruno Haible, Morris Hirsch, Gerhard Huisken, Tom Ilmanen, Allyn Jackson, William Jaco, Arthur Jaffe, Jean-Michel Kantor, Haruta Koseki, Craig Laughton, Bernie Maskit, John McCleary, Bud Mishra, John Morgan, Kunio Murasugi, Eli Passow, Dusan Repovs, Yoav Rieck, Dale Rolfsen, Hyam Rubinstein, Peter Shalen, Zlil Sela, Vin da Silva, Robert Sinclair, Suji Singh, Alex Soifer, John Stallings, Jim Stasheff, Thomas Thickstun, Gang Tian, Anatoly Vershik, Jeffrey Weeks, Gisbert Wüstholz, Shing-Tung Yau, Iwao Yoshioka, Afra Zomorodian.

My greatest and deep-felt thanks go to Christina Sormani from Lehman College and CUNY Graduate Center. Christina painstakingly read two versions of the manuscript and made innumerable corrections and suggestions for improvement to the mathematical details. In countless e-mails she never tired of explaining to me the finer points of Betti numbers, the Ricci flow, the cigar singularity, and many other concepts and procedures that are necessary for the proof of the Poincaré Conjecture. In particular, she came up with the example of the Hydra in chapter 12. Christina is in no way responsible, however, for my characterization of the players in the endeavor to prove the conjecture or for the description of the human drama that unfolded during the verification process of Perelman's work (which is still ongoing). Any remaining errors are, of course, my own.

I would also like to thank my agent Ed Knappman from New England Publishing Associates for his continuous encouragement, and Stephen Morrow at Dutton for his careful editing of the manuscript. My wife, Fortunée, and my children, Sarit, Noam, and Noga, patiently put up with me even while I was suffering slight bouts of Poincaritis. I am very grateful to them.

Index